WOMEN OF THE EUROPEAN UNION

Political and economic changes are re-shaping the employment patterns and social policies that have dominated life in the second half of the twentieth century. Multi-state agreements such as the European Union are intended to foster economic growth, but how will these benefits be distributed? Will these new policies improve the situation of women or create further disadvantages and further responsibilities?

Women of the European Union brings together studies by some of Europe's foremost feminist geographers and sociologists to raise questions about the implications of EU policies for women. Focusing on different scales of analysis, it includes comparative multinational chapters as well as national case studies and in-depth examinations of urban and rural contexts. The book shows how work, family and the state function differently in Spain, Italy, Greece and Portugal than in the countries of northern Europe. Additional perspectives on diversity are provided in chapters addressing discrimination against lesbian women, Thai women married to German men, urban working class women in Denmark, and middle class and suburban couples in the Netherlands.

Women of the European Union challenges gender-blind assessments of the economic and social aspects of the policies of the European Union to examine the implications of union for the diversity of women within member states. We see how place comes together with class, life stage, sexuality and immigrant status to affect ways in which women are constrained and develop strategies to manage their work and daily lives.

Maria Dolors García-Ramon is Professor of Geography, Autonomous University of Barcelona; **Janice Monk** is the Executive Director of the Southwest Institute for Research on Women, University of Arizona.

INTERNATIONAL STUDIES OF
WOMEN AND PLACE
Edited by Janet Momsen, *University of California at Davis*
and Janice Monk, *University of Arizona*

The Routledge series of *International Studies of Women and Place* describes the diversity and complexity of women's experience around the world, working across different geographies to explore the processes which underlie the construction of gender and the life-worlds of women.

Other titles in this series:

FULL CIRCLES
Geographies of women over the life course
Edited by Cindi Katz and Janice Monk

'VIVA'
Women and popular protest in Latin America
Edited by Sarah A. Radcliffe and Sallie Westwood

DIFFERENT PLACES, DIFFERENT VOICES
Gender and development in Africa, Asia and Latin America
Edited by Janet H. Momsen & Vivian Kinnard

SERVICING THE MIDDLE CLASSES
Class, gender and waged domestic labour in contemporary Britain
Nicky Gregson and Michelle Lowe

WOMEN'S VOICES FROM THE RAINFOREST
Janet Gabriel Townsend

GENDER, WORK AND SPACE
Susan Hanson and Geraldine Pratt

WOMEN AND THE ISRAELI OCCUPATION
Edited by Tamar Mayer

FEMINISM/POSTMODERNISM/DEVELOPMENT
Edited by Marianne H. Marchand and Jane L. Parpart

WOMEN OF THE EUROPEAN UNION

The Politics of Work and Daily Life

Edited by
Maria Dolors García-Ramon
and
Janice Monk

London and New York

First published 1996
by Routledge
11 New Fetter Lane, London EC4P 4EE

Simultaneously published in the USA and Canada
by Routledge
29 West 35th Street, New York, NY 10001

Routledge is an International Thomson Publishing Company

Typeset in Baskerville by
J&L Composition Ltd, Filey, North Yorkshire
Printed and bound in Great Britain by
TJ Press (Padstow) Ltd, Padstow, Cornwall

British Library Cataloguing in Publication Data
A catalogue record for this book is available from the British Library

Library of Congress Cataloguing-in-Publication Data
Women of the European Union: the politics of work and daily life/
edited by Maria Dolors García-Ramon and Janice Monk.
p.   cm. — (Routledge international studies of women and place
series)
Includes bibliographical references and index.
1. Women—European Union countries—Social conditions.   2. Women
—Employment—European Union countries.   3. Work and family
—European Union countries.   I. García-Ramon, Maria Dolors.   II. Monk,
Janice J.   III. Series: International studies of women and place.
HQ1590.5.W665   1996
305.4'094—dc20   95–44459
CIP

ISBN 0–415–11879–4
ISBN 0–415–11880–8 (pbk)

CONTENTS

FIGURES

TABLES

ACKNOWLEDGEMENTS

This book began at the International Seminar on 'Women's Work, Employment and Daily Life: A Focus on Southern Europe' held at the Autonomous University of Barcelona in June 1993. Eight of the chapters were originally presented at that time. The symposium was part of a highly rewarding effort supported by the Erasmus Bureau of the European Community and was co-sponsored by the Commission on Gender and Geography of the International Geographical Union. Additional support was received from Institut Català de la Dona, DGICYT (Ministerio de Educación y Ciencia), la Caixa de Catalunya, la Fundació Jaume Bofill, and la Universitat Autònoma de Barcelona. The Erasmus programme has enabled the development of a network of geographers who have introduced gender studies to curricula in European universities. Under its auspices, lively groups of students from the University of Amsterdam, the Autonomous University of Barcelona, Durham University, the National Technical University of Athens, Roskilde University and the University of Sheffield have had the chance to learn together, share viewpoints and widen their understanding. We very much appreciate the enthusiasm of the geographers participating as teachers in this programme, especially Joos Droogleever Fortuijn, Nicky Gregson, Lia Karsten, Kirsten Simonsen, Janet Townsend and Dina Vaiou, who have sustained the network and extended our geographic horizons.

Within the Department of Geography at the Autonomous University of Barcelona, Mireia Baylina, Alba Caballé, Gemma Cànoves, Soledad Morales, Maria Prats, Herminia Pujol, Montserrat Solsona, Antoni Tulla and Núria Valdovinos have collaborated with Maria Dolors García-Ramon to create an unusually 'gender aware' environment that has also made a welcome space for Janice Monk on a number of occasions, as has the Erasmus Group. We very much appreciate their enthusiasm and intellectual stimulation.

For Janice Monk, the project has provided a chance to make a 'home away from home' in Barcelona, and she especially enjoys the continuing hospitality of Maria Dolors García-Ramon, Jaume Torras and Edit

Corominas. It would be hard to imagine a more congenial atmosphere for trans-Atlantic collaboration!

For help in preparing the manuscript, we are indebted to Alba Caballé for the figures in Chapter 1, to Esther Gale at the Southwest Institute for Research on Women at the University of Arizona for seemingly endless revisions and re-formatting of files on the word processor, and to Beth Kangas for research assistance with the literature review for Chapter 1. Tristan Palmer at Routledge, as always, can be relied on to respond promptly, commiserate when necessary, buy the occasional meal or drink and be congenial to middle-aged feminist geographers. We think he has excellent judgement!

NOTES ON CONTRIBUTORS

Isabel Margarida André is Assistant Professor of Geography at the University of Lisbon and also works at the Centre for Geographical Studies. Her research deals with women's employment in Portugal and she is also interested in the comparative position of Portuguese women workers within the European Union, having contributed to the EEC-sponsored project on southern European women and European integration.

Josefina Cruz is Professor of Human Geography in the Department of Human Geography at the University of Seville. She has published widely in the fields of rural geography and population geography and participated in several international programmes on part-time work and rural development in Europe.

Simon Duncan is Lecturer in Social Policy in the Department of Applied Social Studies, University of Bradford and Associate Fellow of the Gender Institute at the London School of Economics. He has carried out research on European housing systems and the local state and is currently working on variations in European gender systems and the uptake of paid work by lone mothers.

Jeanne Fagnani is a geographer and sociologist employed as a researcher at Centre National de la Recherche Scientifique (Paris) and is scientific advisor at Caisse National des Allocations Familiales. She has published on women's journeys to work in urban areas and on the relationships between women's employment and fertility. Her current interests are in the comparative analysis of social and family policies within the European Union.

Joos Droogleever Fortuijn is Lecturer in the Department of Human Geography at the University of Amsterdam. She has published on time geography, the gender division of labour, and the household arrangements of dual-earner families. She is co-ordinator of the Erasmus Network for Geography and Gender and serves on the International Geographical Union Commission on Gender and Geography.

Maria Dolors García-Ramon is Professor of Geography at the Autonomous University of Barcelona. She has published on rural geography, geographical thought, and gender studies and has directed several large comparative regional projects within Spain on the theme of rural women's work and development. She is secretary of the International Geographical Union Commission on Gender and Geography.

Eva Humbeck earned her PhD in geography at Arizona State University. Her research interests focus on gender, cultural identity, and migration in Germany and south-east Asia. At the time of writing she was teaching at Arizona State University.

Eleonore Kofman is Professor of Human Geography in the Department of International Studies at Nottingham Trent University. Her main research interests are gender and political geography, gender and global cities in Europe, and the politics of international migration and national identity with particular reference to France.

Janice Monk is Executive Director of the Southwest Institute for Research on Women and Adjunct Professor of Geography at the University of Arizona. She is interested in comparative approaches to the study of feminist geography, in gender and the landscape, the history of geographical work by women, and approaches to teaching about gender across disciplines. She is vice-chair of the Commission on Gender and Geography of the International Geographical Union.

Ana Sabaté-Martinez is Lecturer in the Department of Human Geography at the Complutense University of Madrid. She has published on rural geography and feminist geography. Her current research interests are in local development, economic restructuring and the gender division of labour in Spain, especially in the clothing industry.

Rosemary Sales is Senior Lecturer in the School of Sociology and Social Policy at Middlesex University, where she is programme leader of the MA in Gender and Society. Her main research interests are in gender, ethnicity and employment. She is currently preparing a book on gender and sectarianism in Northern Ireland.

Jürgen Schmude is Professor of Economic Geography at the Ludwig Maximilians University of Munich. He earned the DPhil in 1987 at the University of Heidelberg with research on the feminization of the teaching profession in Baden-Württemburg. In 1993 he completed his habilitation in Heidelberg with research on spatial disparities in firm formation.

Kirsten Simonsen is Associate Professor of Geography at Roskilde University, Denmark. She is interested in urban studies, social theory in geography, and women in everyday life. She recently published *Byteori og Hyerdagspraksis* ('Urban Theory and Everyday Practice').

Dina Vaiou is Assistant Professor of Planning in the National Technical University of Athens. Her research interests and publications focus on feminist critiques of spatial analysis, the changing features of local labour markets and women's work.

Gill Valentine is Lecturer in Geography at the University of Sheffield where she teaches a course on society and space. She has published several papers on lesbian geographies and is co-editor, with David Bell, of *Mapping Desire: Geographies of Sexualities.* Her other research interests include geographies of food and foodscapes, children's use of space and youth culture.

Paola Vinay is Chair of Prospecta: Cooperativa di Recherche Statistiche e Sociali in Ancona. Her research deals with the informal economy, division of labour within the family, child labour, health promotion, gender and employment, mutual aid organizations, and the reorganization of time schedules of urban services.

1

PLACING WOMEN OF THE EUROPEAN UNION

Janice Monk and Maria Dolors García-Ramon

Political transformation in the face of global economic restructuring is a central feature of life in the late twentieth century. The formation of multistate organizations, struggles for democracy and ethnic rights and challenges to long-standing social contracts are reshaping the worlds of work, family and community relations. The implications of these changes are considerably different for women from those for men, though gender is often ignored or narrowly framed by the politicians and analysts who are creating and writing about new structures such as the European Union (EU) and the North American Free Trade Agreement. Their discourse is predominantly economistic and legalistic, their view of politics gives primacy to the formal, public domain. Feminist scholarship, by contrast, disputes the separation of the economic, political and social, of the 'public' and the 'private'. It understands that the 'productive' economy is interconnected with the reproductive work of daily life and that the 'political' not only relates to the state but permeates relations between women and men in the household and community. In this edited collection we will employ feminist perspectives to explore the changes being experienced in the work and daily lives of women within the member states of the European Union.

The book reflects our own backgrounds and predilections. Feminist scholars today recognize the importance of paying attention to context and of addressing diversity among women on such bases as class and ethnicity. As feminist geographers, we subscribe to those positions, but we are especially sensitive to the importance of place in forming contexts and constructing the experiences of social life. Further, we think it is critical to employ multiple scales of analysis – global, national, regional and local – because different factors are significant in shaping processes of change at each of these levels, while what happens globally can affect the local situation and vice versa. As editors, we have chosen to give more space in this book to the women of southern Europe than is common in most of the collections we have examined. We also want to draw the reader's attention to the fact that the majority of the authors we have included are not native speakers of English and that they usually work outside English-speaking countries. These choices of emphasis and authors reflect our own interest in and experiences of marginality. Geographical work has not been central to feminist scholarship. Feminist perspectives have not been accorded recognition in the geographical literature on Europe. One of us is from southern Europe, the other originally from the southern hemisphere. We are aware of the extent to which masculine and 'northern' ways of seeing the world define concepts, theories and ideas about normative experience and writing, and of the dominance of writing by English-language speakers from Britain and the United States in the literature that frames the questions about and the approaches to understanding women's lives. In multiple ways, we are writing from the periphery, but are striving towards a 'union' that will foster new ways of seeing.

SEEING EUROPEAN WOMEN

When we look at the lives of women in the EU, the picture is quite familiar – they share many of the gender inequalities that are common across the globe: the responsibilities for domestic work that, together with increasing participation in paid labour, result in the burdens of the 'double day'; occupational segregation that keeps them in lower-status positions and a narrow range of jobs; lower income, greater poverty and less public political power than men; and the threats and realities of male violence against women. Yet they also exhibit an array of characteristics that set them apart from women in other parts of the world, including women in other developed regions.[1] By global standards, women in the member states marry later, have fewer children, have relatively high incomes and levels of education, are more likely to live in urban areas, and are more likely to be elderly. By comparison with women in the United States, whom they resemble in many ways, they are more likely to enjoy benefits of the welfare state, such as paid maternity leave and access to state-supported health care, are probably more rooted in place and apparently have more hours free from paid and unpaid work (United Nations 1991: 82).

Much diversity remains among the women of the EU, however. Differences are based in nationality, language, ethnicity, rural, urban or suburban residence, class, age, sexual orientation, marital status and household composition, the nature of gender contracts, the organization of family life, the construction of systems of welfare, and status as an 'insider' (native-born of European heritage) or 'outsider' (immigrant women or the children of immigrants). At this point, rather than taking up such sources of difference systematically, we want to suggest that it is helpful to employ the concept of scale, together with space and time, in seeking to interpret differences; we will illustrate our point by way of some examples.

We begin this discussion by examining women's work experiences at three spatial scales. Though many statistical sources and studies address differences at the national scale, this is not the only scale that needs to be employed. Regional and local data should also be considered. For example, if data on women's economic activity rate at the national level (see Table 1.1) are broken down to the regional level (see Figure 1.1), a very diversified picture emerges that is rich in suggestions. They show, for example, that high ('developed') rates of women's activity can also be found in 'undeveloped' regions in peripheral countries (such as Galicia in Spain and Thrace in Greece), chiefly reflecting the traditional role of women in agriculture throughout their life-course. The north/south contrast commonly noted in national scale studies still holds, but it needs to be nuanced.

Analysis at the local level also reveals that space is a source of difference among women. Figure 1.2 portrays the activity rate of women with two children in the Paris metropolitan area. Districts with a high rate are rare in the centre and are unevenly distributed across the suburbs, although they clearly tend to concentrate in the south. The relocation of jobs in favour of suburban areas and the creation of *villes nouvelles* helped women's integration into the job market, but the availability of public transportation (the SNCF and the RER) and of local social services are also factors that shape the diversity of women's participation and can be identified through local scale analysis (Fagnani and Chauviré 1989), though they might be missed in studies focusing solely on the national level.

In interpreting experiences of daily life, it is critical to pay attention to time – generally a scarce resource for women. Once again, the concept of scale helps us to approach the analysis of time – drawing on daily, weekly and other longer scales. Daily and weekly time are organized in quite diverse ways for women in the member states, as the example of

Table 1.1 Female activity rate in the different states of the European Union, 1991 (percentages)

State	Female Activity %
Denmark	61.1
Portugal	51.6
United Kingdom	51.6
Netherlands	49.1
Germany	47.6
France	46.2
Luxemburg	44.3
Belgium	37.5
Spain	35.5
Italy	35.5
Greece	32.9
Ireland	31.9
European Union	43.4

Source Eurostat (1995) *Regions: Statistical Yearbook 1994*

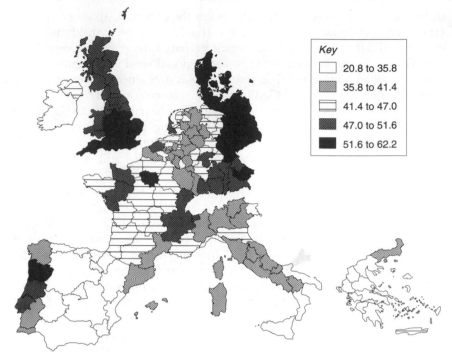

Key
☐ 20.8 to 35.8
▨ 35.8 to 41.4
☰ 41.4 to 47.0
■ 47.0 to 51.6
■ 51.6 to 62.2

Figure 1.1 Female activity rate in the European Union by regions, 1991
(percentages)
Source Eurostat (1995) *Regions: Statistical Yearbook 1994*

school schedules and calendars well illustrates. In effect, in this matter, European unity is far from being a reality: each country has its own traditions and its peculiar way of adapting to social patterns and needs (as well as to its climatic regime). The timetables followed by children in primary education (which strongly condition their mothers' daily lives) differ widely in length (see Figure 1.3). Some countries, like Germany, have a continuous school day, while others, such as Spain, with a three-hour midday break, have a discontinuous one. The reasons underlying these patterns may be quite different, however, so that any modifications will have to be based on different arguments and resources. Greece, Portugal and parts of Italy have shorter days because scarcity of places means that children attend in either morning or afternoon. The same reason does not account for the short hours in Germany. The length and distribution of school holidays also vary greatly among countries (see Figure 1.4), with the shortest holidays in Germany and the longest in France, Greece and Luxemburg. For instance, the French school calendar includes, besides Christmas and Easter breaks, two one-week holiday periods (one in fall and one in winter) though these are distributed in different times in different regions.

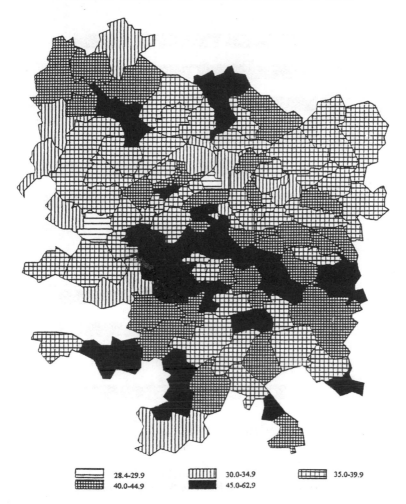

*Married women less than 40 years of age, living with their husband

Figure 1.2 Activity rate of women* with two children (0–16 years old) in Paris metropolitan area, 1982
Source Fagnani and Chauviré 1989

Attention also has to be paid to the longer dimensions of time. In Chapter 13 we look at seasonal time. Here we wish to highlight the scale of generational time. Cross-sectional data, frequently used in national studies, may conceal significant differences between generations. Let us consider the activity rates for Spanish women in 1970 and 1991 as an example (Figure 1.5). In both cases the curve shows a single left-hand peak, which at first glance can be interpreted as reflecting a dominant

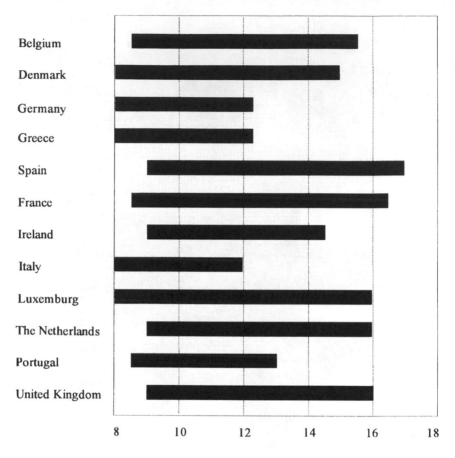

Figure 1.3 Length of school day, first level
Source Eurostat (1992) *Women in the European Community*

homemaker pattern, that is, only young women are in the workforce and rates decline with marriage or childbirth. This interpretation proves to be wrong, however. If data are disaggregated according to generations, a very different picture emerges: between 1970 and 1991, all generations show a pattern of increasing activity after the age of 30 years. This presentation confirms the hypothesis put forward by several scholars that the recent increase in women's activity in Spain is due to married not single women – just the contrary of the initial interpretation. The same is true in other countries and regions where family attitudes and social values have changed dramatically in the recent past. ⌐

Our collection pays particular attention to the importance of scale, partly because of the analytical value of the concept and partly because the realities of the EU are that multiple scales of policy formation and implementation are in effect. While some policies are made at the

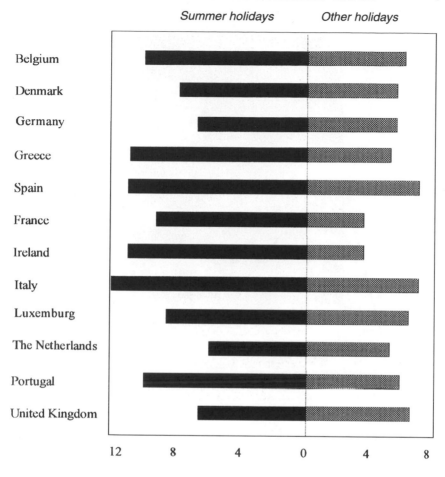

Figure 1.4 Length of school holidays in weeks, first level
Source Eurostat (1992) *Women in the European Community*

transnational Union scale, the EU has adopted a concept of *subsidiarity* which takes the position that for efficiency, policies should be formulated and implemented at the closest possible levels to citizens. Thus national, regional and local agencies remain important even within the transnational European Union.[2]

Our authors work, sometimes comparatively, at these various spatial and temporal scales. They also address an array of sources of difference among women, especially including issues of nationality, ethnicity, class, sexuality, community of residence (rural, urban, suburban), and types of gender contracts and welfare systems. We are conscious that a number of gaps remain. Our authors (like many feminist scholars) are primarily concerned with the lives of women in their younger middle years, who have young children and are in paid employment. Most obvious is the

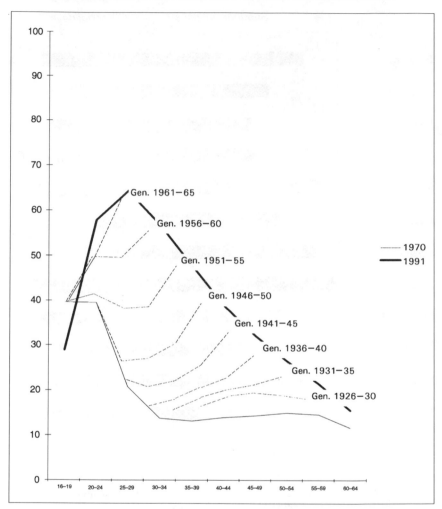

Figure 1.5 Female activity rates by age, Spain, 1970 and 1991
Key Dotted lines represent the evolution of activity rate by generations
Source Solsona, M. and Treviño, R. 1995, p. 198

lack of sustained attention to the lives of elderly women who now account for over 20 per cent of women in most member states, though several chapters do comment on the support given to younger working women by their mothers. Neither do we focus on the lives of girls or of single women – whether divorced, separated, widowed, or never married (other than the chapter addressing lesbian women and aspects of their lives that are affected by prohibitions against marriage). Given the growth of new forms of households, such diversity is important to address. Additionally, diversity relating to religion, language and ethnicity, both within national units such as Spain and Belgium and across national units (for example, Muslim women in Britain, France and Germany) and the different experiences of migrant women that reflect their places of origin and entry status require much more attention, as do questions that connect place with women's personal and cultural identities. We would also like to see more research on women's multiple relationships beyond the family – with co-workers and in the community, and on aspects of their life other than work, such as health and leisure. Our hope is that we open some avenues for viewing the complexities of the politics of work and daily life for women in the European Union and encourage others to take up questions we have not explored here to the extent that we would have liked.

HAZARDING COMPARISONS

Scholarship that attempts to cut across national boundaries faces an array of hazards. One critical issue is the diversity of languages in which relevant literature exists. In our case, the individual authors have drawn on work in the languages of the places they are studying, but it is also evident that English-language sources predominate, particularly where comparative and theoretical works are cited. Thus the range of positions and interpretations that might inform understanding becomes narrowed. Then too, the EU itself (and its predecessor bodies) is not a stable unit. When we began this project, the Union was called the Community, and the member states numbered twelve. Since that time, three more countries (Austria, Finland and Sweden) have joined. Some of our cross-national chapters draw examples from countries in the larger set, but our national, regional and local scale studies draw on the earlier grouping and, even then, it has not been possible to include detailed chapters on each of the member states.

Another major difficulty lies in the lack of comparability of empirical, especially statistical, data. What gets counted, how categories are defined, what time periods are covered, and what degree of detail is presented in reports: all these factors differ from country to country. For example, if we wish to address the question of the varying nature and degrees of occupational segregation across member states, we are hampered by the lack of standardization of their systems of occupational classification (see Delacourt and Zighera 1992: 143–7). Similarly, if we are interested in the nature and extent of women's educational opportunities and attainments, we are faced with making sense of different national laws regulating ages of leaving school, varying kinds of transitions between levels within educational systems, differing emphases on initial and in-service vocational training and so on (see Delacourt and Zighera 1992: 57–66). Compounding these problems is the failure of almost all reporting systems to pay any attention to women's reproductive work. Further, coverage of such concerns as values and attitudes is patchy and the tendency is to conduct such studies at the national scale as, for example, the periodic Eurobarometer surveys, and not to examine women's opinions separately from men's (see, for example, Commission of the European Communities 1986). As Jan Windebank (1992) has argued, reliance on the official, national level data compiled in European Community (Union) publications may very well have led to an over-reliance on factors relating to the state as the source of differences among women. Commissioning cross-national studies to address a specific research theme is also not without hazards, as authors have difficulty coming to consensus on the appropriateness of methodologies or definitions for their own contexts (Millar 1992).[3]

Such considerations point to the need to combine larger-scale transnational and cross-national research with in-depth regional and local case studies and to employ qualitative as well as quantitative research methods. Only in this way will we gain a better sense of the processes that are operating to shape women's work and daily lives in specific contexts, to represent women as agents reflecting on and shaping their own lives, to portray the diversity of women's experiences, and to identify levels at which action for change needs to be initiated. We have aimed to bring together such a mix of scales and methodologies, and see their integration as a strength of this volume.

SITUATING THIS BOOK

Europe became a subject of growing importance for geographers with the enlargements of the European Economic Community in the 1970s and 1980s and the advent of the EU after the Maastricht Treaty. Feminist scholars, likewise, have turned a significant amount of attention to the meanings of the 'New Europe' for women. We thus consider it important to situate this book within that literature, mostly published in the 1990s, though some works also appeared in the 1980s.

First we will look at a sample of the geographical textbooks on Europe, written in different languages. Reflecting their disciplinary orientation, they are often sensitive to differences at regional and even local spatial scales, but they are almost completely blind to gender. This is true of otherwise generally good introductions to Europe such as Williams (1987), Ilbery (1990), Minshull (1990) and Dawson (1993). The same can be said of French, Italian and Spanish books, such as those by Berthaud (1991), Migliorini and Pagliettini (1993), Arroyo (1988) and Villanueva (1994). A few give spotty attention to women or gender when they refer to particular issues such as work and unemployment, equal pay, new forms of family life, demography, social funds and the like, but gender is not integrated in any meaningful way through their works. This is the case for Pinder (1990), Clout et al. (1989), Williams (1991), Masser et al. (1992), Cole and Cole (1993) and Blacksell and Williams (1994). These findings are not surprising if we keep in mind that the gender approach in geography developed in the late 1970s and early

1980s and that new trends in research always take their time to percolate into general works and textbooks.

A slightly different view emerges when we turn our attention to geographical works that retain a broad European scope but present more specialized research. For example, in the edited work by Noin and Woods (1993) on population, five of eighteen chapters make explicit and abundant references to gender and another is devoted entirely to the role of women in post-industrial economies. In the collection edited by Garofoli (1992) on endogenous development, one chapter deals centrally with the critical implications of women's work in small enterprises for the development of Greek intermediate regions. The same is true for the collection edited by Hadjimichalis and Sadler (1995). Scattered through Scardigli's (1987) treatment of work and of diversity and unity in European ways of life are comments about women's work and the diversity of European gender contracts. Books that include a more anthropologically oriented approach such as the works by Todd (1990) and Woolf (1993) centre discussion on differences in family structures and therefore implicitly point to gender relations as crucial in explaining differences between northern and southern Europe.

Nevertheless, it is really only in recently appearing special issues of research journals on Europe that sustained attention to gender perspectives can be found. In France, for example, Jeanne Fagnani and Jacqueline Coutras guest-edited an issue of *Espace, Populations, Sociétés* (1989) devoted to the theme of sex and space in which six of ten empirical contributions are on European women. *Iberian Studies* (1991) organized an issue (edited by Maria Dolors García-Ramon) around the topic of women, space and society in Spain and Portugal and the Italian journal *Inchiesta* (1992) focused on women in the southern areas of South European countries. Recent issues of the Swiss journal *Geographica Helvetica* (1994) (in German) and the German journal *Geographische Rundschau* (1995) incorporate studies of European women, with the latter addressing the rarely discussed topics of older women and women living alone. Two special issues of *Environment and Planning A* (1994) edited by Simon Duncan took up theoretical and empirical aspects of diverse forms of patriarchy in Europe and in Spain; a special issue of *Documents d'Anàlisi Geogràfica* (1995) edited by Montserrat Solsona focused on women's work and daily lives in Europe with an emphasis on the southern countries. The rising level of research and interest is thus clear, though the fact remains that in the new journal *European Urban and Regional Studies* four issues were published before any gender-aware work appeared (1995). The geographical range within Europe of the special issues is quite wide, but the scale of analysis is primarily local and sometimes regional. Relatively little cross-national work appears. Thus, our book is unique in being prepared for a wider audience than can be reached by the only currently available geographical works which highlight women and gender in (western) Europe and we go beyond their orientations by cutting across geographical scales.

In addition to the geographical literature, a considerable body of multilingual feminist works that now exists across disciplines addresses contemporary women's work and daily lives in Europe. It is quite beyond the scope of this introduction (and our linguistic capacity) to survey the entire field, particularly all the local and national studies. Rather, we will limit our comments to works that address the EU (and its precursors) directly, or that are transnational, multinational and/or comparative across national boundaries, since these speak most closely to our project.

The nature of EU policies directly referring to women and the politics of their development have been well documented, so that we need not rehearse the details. Among the works we have found especially useful are Buckley and Anderson (1988), Cunningham (1992), Duchen (1992), Hoskyns (1992), Mazey (1988), Meehan (1992), Springer (1992) and Vallance and Davies (1986). Basically, these authors analyse the origins and implications of policies about equality in the workplace that have been created since Article 119 of the Treaty of Rome (1957) stipulated that men and women should receive equal pay for equal work. They review the subsequent Directives – European Community legislation which is binding on member states, but which each may implement in its own way – that extend to equal pay for work of equal value, equal treatment at work and equal treatment in social security. They also outline the dimensions of Action Programmes on the Promotion of Equal Opportunity for Women since the 1980s, which have sponsored research on specific groups of women (for example, single women) or issues (for example, childcare) and formed the basis for generally small-scale and short-term educational and training programmes in the member states.

Substantial agreement exists among these feminist scholars about the limitations and effectiveness of the Directives and the Action Programmes, as well as about the roles of the European Parliament and the European Court of Justice in shaping women's lives. First is the narrowness of the focus on women as workers, rather than as citizens, so that many aspects of women's daily lives, including those aspects of daily life that affect experiences of employment, have been excluded from consideration and/or policy and action. The public-private connections, well established in feminist scholarship, are not evident in EU Directives. As Catherine Hoskyns notes, citing the Belgian trade unionist Emilienne Brunfaut, 'these are men's rules for women's rights' (1992: 22). Second is the resistance of member states or the low priority accorded by them to creating or implementing policies that would give effect to the Directives, though it is also apparent that the existence of the European policies has helped to keep women's issues on the agenda and that national provisions adopted to comply with European obligations are harder to repeal than those of purely national origin (Hoskyns 1988).

A third commonly identified theme is the EU's treatment of gender separately from other forms of inequality. In particular, issues of racial discrimination are not incorporated into policies dealing with equality between women and men, so that the specific problems of minority women are masked, as are the situations of those with disabilities (Cunningham 1992). Nevertheless, the research supported under the Action Programmes has paid attention to diversity among women, addressing, for example, older women (Coopmans et al. 1989), 'solo' women (Millar 1991) and migrant women (Prondzynski 1989).

Finally, in discussing the progress of policy development and implementation, the authors note the importance of changing temporal contexts; in particular, the coincidence of 'second wave' feminism with the heightened activities around equality issues in the 1970s and the dampening effects of economic recessions and increasing political conservatism since the 1980s. All these considerations shape assessments that the transnational policies have been of limited significance in bringing about any reduction of vertical or horizontal gender segregation in employment, though this is not to conclude that they have been without value in enhancing and, to some extent, sustaining policies directed against discrimination, if not promoting equity. Looking to the future, optimism for women is very guarded. Since the goals of the Single European Market are to enhance the economic competitiveness of member states, the disadvantaged economic position of women could well worsen as labour laws are weakened and emphasis put on cost-cutting. While women's employment is likely to continue growing, especially in the service sector, growth will probably occur in jobs that are insecure and low-paying. Further, given the implications of the gender division of domestic responsibilities, women are seen as less likely to benefit from policies to support the 'free movement' of labour (Springer 1992).

Another valuable strand of literature presents cross-national comparisons of aspects of women's lives, especially of employment. We note the descriptive and statistical publications such as *Women in the European Community* (Delacourt and Zighera 1992) and Meulders et al. (1993) which provide extensive demographic, economic and social data in tabular and graphical form. We also find useful cross-national studies of gender relations and employment, such as the comparative work on Britain and France by Crompton et al. (1990), the twelve-country survey of retrospective data on family and work history by Kempeneers and Lelièvre (1993), collections such as Elson and Pearson's *Women's Employment and Multinationals in Europe* (1989), and special issues of *Women's Studies International Forum* (1992) and *Feminist Review* (1991) that also include theoretical, bibliographical, and case study articles. Almost all of these works, however, confine their analyses to the national scale, though to some extent they consider social diversity (for example, Morokvasic 1991 on migrant women). Explanations that reflect the role of the state are most prevalent – a phenomenon that Windebank (1992) sees as an artifact of the scale of analysis and the nature of the data. Differences (or similarities) among women that might reflect processes or variables that are more significant or more readily investigated at other scales (regional or local) are thus not highlighted. Further, perhaps because of their later entry into the European Community, women in the southern countries of Spain, Portugal and Greece receive less attention than women of the larger, older member states (Britain, France, Germany) or the eastern European women in post-Communist states where recent change has been dramatic.

Feminist scholars have also devoted considerable attention to the study of women's movements and women's public political activity within Europe, including comparisons within Europe and between Europe and the United States (see, for example, Jenson 1985; Kaplan 1992; Katzenstein and Mueller 1987; Lovenduski 1986). Other pieces have focused briefly on specific women's and feminist groups such as the International Gay and Lesbian Association (Power 1991); the European Forum of Socialist Feminists (Cockburn 1992) and the European Women's Lobby (de Groote 1992; Hoskyns 1991). The more in-depth studies highlight the importance of cultural and historical contexts in shaping the directions of women's movements in different countries, as well as their relations to state politics and other political actors such as trade union groups. From our perspective, the recognition that the degree of centralization or decentralization of the state is salient in assessing the impact of feminist movements on public policy is an important reason for fostering research at the subnational scale (Katzenstein and Mueller 1987; Lovenduski 1986). We also find of considerable interest Hoskyns's observations (1986) that European women's movements have pursued concerns (such as domestic violence, reproductive rights and the need for rape crisis centres) other than the issues of formal equality in employment that have dominated the transnational policy-making. Her comments take us back to the position of defining the 'political' in women's terms and of the need to connect work and 'daily life'.

CONSTRUCTING THE MOSAIC: WOMEN'S WORK AND DAILY LIFE IN THE EUROPEAN UNION

We begin this volume with four chapters that look at member states within the transnational context of the EU, emphasizing the development of conceptual frameworks for comparative analysis and the politics of pursuing difference or convergence. Chapter 2, by Eleonore Kofman and Rosemary Sales, takes up the theme of welfare regimes, thus highlighting the social as well as the economic aspects of policies. They criticize the best-known comparative studies for being insufficiently gender-aware or for being too narrow in their definitions of welfare; these studies tend to include only those elements of life that are quantifiable through some form of market exchange and focus primarily on policies that regulate the labour market. Kofman and Sales stress the need to broaden the definition of welfare to include its less quantifiable aspects such as support and care for people's physical, psychic and emotional well-being. Drawing on Connell's (1990) analysis of 'gender regimes', they suggest how the diversity of European welfare regimes might be better understood through a theoretical framework that takes account of gendered state structures, relations and interventions. They emphasize the importance of creating a framework to understand diversity because, despite tendencies to convergence, distinct welfare regimes remain with important implications for gender relations.

The concept of integration and of an emerging 'New Europe' – as projected through the Maastricht Treaty – is at the heart of Chapter 3 by Dina Vaiou. In this process of integration, many of the European Union policies favour the most advanced and dynamic parts of the Union while marginalization and poverty continue to be the reality for many regions and social groups. Indeed, Vaiou argues, the experiences and priorities of the stronger components are represented as normative, and weaker groups are constructed as 'other'. A good example of such a construction is the situation of women of the countries of southern Europe. Within this framework, Vaiou analyses the experience of women's paid and unpaid work, and the importance of the family as a 'protective net' which, to a certain extent, substitutes for the well-developed welfare systems of most of the countries of the north. Her findings demonstrate that 'Fortress Europe' already contains a lot of what it seeks to leave out of its borders and that it continues to create social and geographical peripheries in the process of its own construction.

Like Kofman and Sales, Simon Duncan (Chapter 4) takes up the task of developing a more satisfying framework for comparative analysis. His focus is on gender relations and inequalities. He criticizes existing models for their inability to capture the processes leading to the observed patterns and especially for their neglect of spheres of life other than paid work. Building on the notion of the 'gender contract' and using empirical indicators of gender divisions of labour, political motherhood and the political family, Duncan explores the differentiation of patriarchy in western Europe. Countries with an 'equality contract' (Denmark and Sweden) are distinguished from those with a 'dual role contract' (Finland and France), a 'housewife contract' (notably Germany) and 'private patriarchy' (Greece), although transitional forms are also acknowledged (for example, Norway, Spain and the United Kingdom as well as the former German Democratic Republic). Supporting our contention that multiple scales of analysis are important, Duncan then argues that gender relations are not always, or indeed mostly, defined at the level of the nation-state, and demonstrates his point with regional scale studies of Britain and Germany.

Chapter 5, by Gill Valentine, raises the question of when it is most strategically useful politically to maximize transnational similarities and concerns across the EU and when to pay greater attention to national differences. She directs her question to the struggles of lesbians facing discrimination in the workplace, beginning with an examination of lesbians' shared experiences across nations. Since a key aspect of the EU is to facilitate the free movement of workers, she asks how the differences in legislation among member states with regard to discrimination and the rights of citizens might impinge on the free movement of lesbian workers. Lesbians could build transnational solidarity to challenge discrimination. Yet her chapter also reveals that substantial differences exist in forms of oppression at national, regional and local levels and in ways in which other identities intersect with sexual identity in various contexts. Thus she argues for fluid strategies, allowing the tide of communion to ebb and flow as seems most effective.

Jeanne Fagnani also pursues the theme of the implications of policies for the conjunctions of private life and employment, highlighting the ways in which values shape national policies with respect to the family. Chapter 6 is an examination of the specific patterns of women's employment and fertility in France and the former West Germany, focusing on the interactions between familial and professional spheres. In both countries the completed fertility rate had begun to decline long before the Second World War. However, for women born in 1955, the completed fertility rate is still much higher in France than in West Germany, even though French mothers are more often economically active than German ones. One of the important explanations lies in the type of family policies of the two countries which play an important role in creating a more-or-less guilt-induced environment for mothers who want to be economically active. While in West Germany family policy confirms and strengthens the antagonism between maternity and employment, in France the model of the 'working mother' (as in Sweden and Denmark) is fully integrated into the family policy.

The importance of state level analysis and of questioning models developed from 'northern' experience is further developed by Isabel Margarida André in Chapter 7. Portuguese women exhibit one of the highest rates of economic activity in the European Union. Part-time jobs are unusual, women's access to central segments of the labour market is relatively large and, among young people, women show a higher level of education than men. These phenomena can be understood in relation to Portugal's recent history and economic development – the Colonial War (1961–74), the emigration of the 1960s, the revolution (1974) and the entry into the European Community in 1986. Yet Portuguese women continue to suffer some forms of occupational segregation and an array of social institutions, such as trade unions, have marginalized concerns about gender equity.

Jürgen Schmude also brings time and space together in his examination of women's occupational segregation in Germany since 1945, focusing on variations within Germany. Because employment, training and childcare policies contrasted markedly between the separated states of the Federal Republic of Germany and the German Democratic Republic, he first considers these regions separately then explores how the processes of unification have begun to affect women's employment. His study shows how occupational segregation persists, though in changing forms in different locations and times. His analysis of regional differences (between *Länder*) within the two Germanies reveals how the growth of service sector employment has contributed to diminishing some spatial distinctions in women's rates of employment.

A different approach to change within Germany is taken by Eva Humbeck. As the EU legally promotes the movement of citizens of member states across internal barriers, fostering 'transnational' perspectives, barriers are being raised against outsiders, especially those from Third World countries. In this context, Humbeck focuses on one group of outsiders, Thai women married to German men. These women are commonly represented as participants in or victims of the international sex trade, but the chapter presents a more complex view, looking at the women's agency and values. She contrasts constructions of immigrant identity being forged by the German governments, non-governmental organizations (NGOs) and the Thai women. While official policies call for 'integration', as do some conservative NGOs, other German groups envision a multicultural society, and the women seem to prefer a more flexible model which values the economic security and family life they have established in Germany but sustains their emotional ties to Thai culture. Her work suggests the need for further research on transnational and bi-cultural issues within the EU (and global) context.

Questions of values and the family are also central in the chapter by Paola Vinay. In the 1960s and 1970s, the prevailing cultural model for women's work in Italy placed priority on the family. But over the last fifteen years, important changes have occurred in the demographic structure of the family, in women's access to higher education and in their attitudes to paid work. A new cultural model is emerging in which women do not want to choose between work and family. The desire of women to live all the dimensions of life is at odds with the prevailing rigid organization of work schedules and the gender divisions of work roles, however. These circumstances underlie the current debate in Italy about the reorganization of time schedules in work and daily life. It centres on the growing need in post-industrial societies for greater flexibility and on the acknowledgement of the social value of caring work within the family, and for a more equitable gender division of that work if full and equal participation in all dimensions of life is to be enjoyed by all. Vinay integrates into her discussion the theme of regional and local differences in changing patterns of work and examples of local experiments in developing policies to reorganize daily time.

Joos Droogleever Fortuijn situates her study of the integration of paid work into everyday life at the intra-urban scale within metropolitan Amsterdam. This allows her to focus on changing conjunctions between context (city and suburb), household type and the life-course. She looks at fifty dual-earner families living in three different districts – the inner city, the inner suburbs and a new town area. Her findings show that, although the rise of dual-earner families has occurred across the country, the gender contract represented in the arrangements that households make as they integrate paid work into daily life is clearly context- and class-specific. Within the possibilities and limits of the housing market, households shift over the life-course between urban and suburban contexts to secure a supportive environment.

A still finer-grained interpretation of the integration of work and daily life is provided by Kirsten Simonsen who addresses the development of the mode of life of women and their families over the last fifty years in a working-class neighbourhood in Copenhagen. Using life-history methodology and considering the interplay between these individual biographies and the larger social structures, she explores areas of continuity and change in the family and the neighbourhood – in social interaction and in attitudes towards work. The findings challenge two widely held generalizations: first, the general understanding of modern urban life as totally disembedded from attachment to place; second, the existence of close family and kinship networks as a phenomenon of southern but not northern European societies. Her analysis suggests the persistence of these ties (at least among the working-class families she has studied) even in the cities of a society as penetrated by welfare institutions as the Danish one, although the practical importance of family and kin in the management of daily life has been modified. Her work reminds us of the need to study what endures, as well as what changes, as policies and values shift.

The importance of contextually sensitive studies of the welfare state and its relations with the family is reinforced in the chapter by Maria Dolors García-Ramon and Josefina Cruz. They conduct research at the regional scale on the consequences of policies for the daily life and work of women agricultural day-labourers, a conspicuous group in southern European countries, paying attention to the ways in which women manipulate the resources emanating from state programmes. In comparison with northern Europe, the welfare state has been both late being established and somewhat undeveloped in southern Europe. Public services that can be viewed as indirect subsidies for low-income people are scarce or lacking in Spain, particularly in rural and economically poorer regions such as Andalusia where the research was conducted. Nevertheless, the welfare state is present in the rural economy of these areas through policies of unemployment subsidies for casual day-labourers. These have made women who are casual workers in agriculture more visible, but have also indirectly devalued women's work, as women more often than men tend to emphasize work in the subsidized sector, often in trivial public works organized at the local state level.

In the final chapter, Ana Sabaté-Martinez connects local conditions to global processes, seeking understanding of the profound changes that are occurring in the work and daily lives of rural women in central Spain. Here, as in other Mediterranean countries of Europe, the agricultural crisis is reducing incomes on family farms and families are compelled to diversify their activities and seek non-farm work. In this context, processes of decentralization associated with strategies such as subcontracting and development of the informal economy in such industries as electronic components, food-processing, leather, footwear, clothing and textile manufacture are drawing rural women into industrial work. They offer important advantages to employers seeking to cut production costs through women's willingness to take temporary, 'unskilled', informal and poorly paid jobs. Central to understanding this situation are the gender politics within the family and the social devaluation of women's work.

CONTEXTUALIZING DIFFERENCES AMONG WOMEN OF THE EUROPEAN UNION: PLACE, WORK AND DAILY LIFE

In interpreting women's work and daily lives, our authors' close attention to context highlights the multiple ways in which place is significant. While most obviously it is important in accounting for different experiences among women, no less is it implicated in shaping the discourses of scholars and policy-makers and of the women themselves — about what they value and consider possible. Most clearly, the differences are evident in the *conceptualizations of 'work'*. What does 'work' include? Who is 'active'? What is considered 'typical' or 'atypical'? What is 'full-time' or 'part-time'? Whose assumptions are dominant when we move into international comparative study or policy-making beyond the local scale? Whose interests are served by varying conceptualizations? Vaiou provides a sustained discussion of such questions when she argues that most research, official documents and policy-making in the European Union assume the model of lifelong, full-time, unionized employment for a family wage in big workplaces in industry or the services. In this model she sees not only a male (and historical) bias, but also a 'northern' bias. Most women workers in the 'south' would be considered 'non-active' or 'atypical', clustered as they are in very small family firms, in the informal sector, as casual workers or as family workers (quite often unpaid) and seldom unionized.

When work is examined at the national scale, the extent to which women are engaged 'part-time' varies widely, as Duncan points out: over 60 per cent of all employed women in the Netherlands work part-time compared with around 10 per cent in Greece, Portugal and Spain. Similarly, engagement in full- versus part-time work differed greatly among women of the former East and West Germanies and these differences have been magnified since reunification as older women in the East have suffered most from job loss (see Chapter 8).

To some extent, the apparent differences reflect the criteria used to define 'part-time' (and whether it is additionally identified as 'short' or 'long' part-time), but even if statistical similarities exist, the reasons why women are working part-time, which women are involved, and the implications of that work for their daily lives and for career advancement differ markedly. Commonly, their choices reflect the impact of state policies with respect to the provision of childcare (see especially Chapter 6), but even with extensive state-supported care, as in Denmark (see Chapter 12), high proportions of women can be part-time workers. Which mothers work part-time and how employment is combined with other activities also vary among places within a nation, as Droogleever Fortuijn reveals – different residential locations within the Amsterdam metropolitan region offer different possibilities for employment and services and also attract people whose career aspirations, economic capacities and life-style preferences lead to alternative ways of combining hours of work and family life. Further, it is not necessarily mothers who work part-time. In the former East Germany, older women have been more strongly represented among part-timers, with such work facilitating their transition to retirement (Schmude). Also, as André notes with respect to Portugal, part-time work may mostly involve older women when they are concentrated in traditional sectors of the economy (agriculture and domestic work). Indeed, examination of the working lives of rural women reminds us that part-time work also needs to be understood as seasonal work, not only part-day employment, and that the particularities of the local economy, whether in rural manufacturing, as discussed by Sabaté-Martinez, or in agriculture, as examined by García-Ramon and Cruz, condition which women work, when and how much.

Women's involvement in full-time work also has different explanations in different places. As Fagnani indicates, state policies in France are crucial in supporting high rates of participation in full-time employment by diverse groups of women, including those who are married with more than one child. By comparison, in southern European countries it is the female family network that supports the higher recorded rates of participation in full-time work as André and Sabaté-Martinez make clear in their discussions about Portugal and Spain. Although households may appear to have made a transition from extended to nuclear family groupings, residential proximity of mothers, mothers-in-law and sisters remains central in sustaining close family relations and support for employed women. André makes the additional point that the need for income among Portuguese families is such that the choice of part-time work is not attractive.

In treating 'daily life', our authors focus a great deal on social policies and welfare. Once again, questions of discourse and scale of analysis are important. Kofman and Sales outline the multiple approaches to social policies that exist in contemporary Europe – those that are oriented towards income maintenance for individuals unable to support themselves through the market or the family (such as unemployment benefits), those that provide public services (such as education and childcare) and those that regulate working conditions (such as limiting hours). Their discussions of the typologies of welfare systems that have been developed highlight differences at the national scale, which are certainly of major significance. They also take up the importance of the local scale, especially in southern Europe, where the nature and extent of welfare provision can be seriously misinterpreted if only national level analysis is employed. Two chapters reinforce this point. In the depressed agricultural region of Andalusia, where the provision of social services is very weak, but where the regional Socialist government provides transfer payments, these policies are critical for the survival of seasonal workers (see García-Ramon and Cruz). Analysis that focused only on the national level or on the provision of services would miss this major governmental contribution to welfare. Vinay also indicates that the Italian state often chooses to provide financial assistance to families rather than social services and that there is a lack of a clear national policy. Provision differs greatly from region to region and from municipality to municipality, with the differences being exacerbated as the national government has made cuts in commitments to welfare expenditure. Social and health services are concentrated in the larger municipalities of the northern and central regions, while they are almost absent from the smaller ones, especially in the inner mountain regions. The budget of a municipality and the political orientations of its government thus become of prime importance in determining the support for children, the elderly and non-self-sufficient people.

Also important in the discourse around welfare is the question of which aspects of life and which people are considered by scholars and policy-makers to fall within this domain. Kofman and Sales also pursue this point. They show how the focus on workers' rights leaves many, such as the elderly, outside the domain of unemployment discussions in the EU. They also raise an array of issues, from elder care to questions of violence against women and the rights of minority and immigrant women, which are beyond the bounds of the more traditional discussions of women and welfare which emphasize childcare and maternity issues.

Not surprisingly, discussions of the *gender segmentation of the labour force, gender-based inequalities in income and the extent and nature of discrimination* recur throughout our book. All are widespread across the member nations of the EU, with the broad patterns being outlined by Simon Duncan. But again, we see how place (and the interaction of place and time) constructs the ways in which women experience these processes and outcomes. Schmud's analysis of the historical and geographical patterns of feminization of the teaching profession reveals significant rural and urban place-based differences; André shows spatial and temporal variations within Portugal in women's access to education and professional work as larger national and international forces (war, emigration and revolution) impacted on their lives; and Duncan presents evidence that within Britain, whether women have dual roles as workers and homemakers or are predominantly homemakers varies quite substantially between areas of the country, as does the extent to which a gap exists between male and female wages.

Specific groups of women experience discrimination in ways that reflect their personal characteristics and how these are evaluated by governments, employers, other members of society and their families. At the transnational scale, Valentine's chapter on anti-lesbian discrimination in the EU addresses the importance across countries in attempts to resist discrimination, but also points out how national level policy differences (for example, with respect to immigration and citizenship rights) have a bearing on lesbians' freedom to seek work and residence in different places. She further addresses place-based differences in lesbians' freedom to express their sexual identities by comparing conditions in Greece, Italy and the Czech Republic, asking how expansion of the Union to include eastern European countries might complicate efforts to mobilize against sexual discrimination. Sabaté-Martinez also brings the transnational and the local together in showing the vulnerability to exploitation of lower-skilled, female workers in the rural areas of southern Europe who, like women in certain peripheral cities and countries of the Third World, are seen as a 'docile' labour pool to be used in low-paying, repetitive, casual and home-based manufacturing production.

Across places, women facing such kinds of discrimination have not been receiving significant or effective support from political and labour institutions. Though some actions and policies instigated by the EU have been directed towards the reduction of inequities in the workplace and against pay discrimination among member countries, Valentine demonstrates that they have not been supportive of efforts to encompass sexual preference within this domain. Additionally, André and García-Ramon and Cruz point out that although the activities of trade unions and leftist organizations have had the effect of improving working conditions and income for women (as well as men) – for example, by introducing collective bargaining for wages – they have been disinterested in addressing occupational segregation, the feminization of tasks, or the greater susceptibility of women to unemployment. Thus women continue to be disadvantaged in the labour market. Similarly, Kofman and Sales indicate that the introduction at the national level of state policies that have benefited women workers, such as Swedish taxation and parental leave policies, has not counteracted the gendered division of labour.

Indeed, they identify Sweden as still having one of the most horizontally and vertically segregated patterns of employment in Europe. Where some diminution of traditional patterns of gender segregation is recognized, it appears to be more a function of women's gaining access to higher levels of education and a greater diversity of training programmes (see Chapters 7 by André and 8 by Schmude), though these advances are precarious and most commonly unidirectional (with women moving into previously male occupations rather than vice versa).

Inequalities also persist across the EU in women's disproportionately larger share of responsibilities for household work and the care of children and the elderly. All the chapters address the difficulties women face in *managing the 'double day'*. Nevertheless, they highlight quite different patterns, especially between countries of the north and south, in women's strategies for coping with their dual roles. Vaiou notes the persistence in southern Europe of prescriptive behaviours, rights and duties that establish the family as a protective net, especially in times of unemployment, and which renders services to children, the sick, old and disabled. Within the family, it is women who form the basis of this net. Her contentions are supported by the chapters on Portugal, Spain and Italy.

State support for child-bearing (paid maternity leave) and child-rearing (parental leave and childcare services) is more common in the countries of the north, though here the nature and degree of provision vary widely, with quite different consequences for women's fertility, type of labour force participation, and advancement in the workplace (see chapters by Duncan, Fagnani, Schmude, and Kofman and Sales). Women in these societies, however, do not benefit equally or make the same choices, and again scale is salient in shaping differences in experience. Duncan elaborates on differences among local and regional governments in childcare provision within Britain; Simonsen describes the persistence of family support in a relatively stable working-class neighbourhood of Copenhagen, despite the complementary support of public services, though she also notes the development of new neighbourhood networks that have their origins in connections established through the public institutions serving children. Droogleever Fortuijn contrasts the strategies of dual-earner families in three Dutch neighbourhoods where the absence of both state services and the protective family network places on couples dealing with the double burden the responsibility of generating ways that are within their financial means (thus leading to class differences in ways of managing the double burden) and that reflect the variations in local urban infrastructure.

More innovative approaches to addressing the pressures of combining paid work and domestic responsibilities are evident in recent efforts in Italy to create new organizations of time. Vinay reports on the debate that has focused on the revision of the schedules of all urban public and private services so that they are compatible with work schedules in a post-industrial society. The experiments in revising schedules are, however, legislated at the local, municipal level, so that they affect women unevenly across place.

Distinctive cultural and personal values underlie the differences in discourse, the persistence of discrimination and occupational segregation and the burdens of the double day that we have identified; they also shape the policies and private behaviours that attempt to transform work and daily life. The chapters by Duncan and by Kofman and Sales present extended discussions of scholarly attempts to model the cultural assumptions that affect differences in gender relations, 'gender contracts', and welfare state systems across the nations of the EU. They demonstrate clearly the difficulties of creating gender-sensitive, *analytical* models that are, on the one hand, sufficiently inclusive of both work and welfare and, on the other hand, sensitive to the importance of geographical scale. Most of the well-known models, for example, are elaborated only at the national scale, thus being insensitive to deeply rooted cultural and political values that operate on local levels.

Again, we wish to highlight that our chapters point to the value of *multiple* scales of analysis for developing understandings of differences. Thus, at the national scale, Jeanne Fagnani's comparison of French and (West) German family policies (with respect to maternity and parental leave and childcare support) provides new insights into relationships between women's engagement in paid work and their rates of fertility. She demonstrates reciprocal interactions between public opinion, state policy and individual choices which operate to create higher fertility *and* higher rates of participation in full-time employment among French women in comparison with both lower fertility and lower workforce participation (and more part-time work) among (West) German women. Her work challenges the belief that direct causal relationships exist between the declines in the birth rate and women's increasing labour force participation.

Similarly, analyses at the national scale of policies on marriage, sexuality, divorce, immigration and citizenship rights provide important insights into the barriers against which women must struggle in order to gain greater autonomy and extend their range of choices. Valentine identifies marked differences in the protection afforded to lesbian workers under the anti-discrimination laws between member states of the EU as well as in the legal status of unmarried couples, specifically of gay and lesbian partners. For example, in Denmark, registered partnership laws confer on same-sex relationships similar rights to those enjoyed by heterosexual couples through marriage, and recent Spanish legislation permits the registration of lesbian and gay couples, but the existence and nature of such regulations are quite uneven across countries. In the context of the free movement of workers promoted by the EU, such differences have important consequences. Thus a Danish lesbian would face the prospect of separation from her partner if she migrated to the

UK for work, but her British counterpart would not be similarly disadvantaged if she wished to move to Denmark. In her chapter, Humbeck notes that Thai women are only partially protected by German law, even when married to German citizens. They stand to lose residency in the case of divorce within the first three years of marriage, and German courts generally grant child custody to the German father. Legal suits against physical and psychological abuse have been difficult for them – basically the laws are designed to give priority of rights to the native-born over the immigrant

Local, household and individual scales of analysis are, however, especially valuable for understanding aspects of choices with regard to work and managing everyday life. Droogleever Fortuijn's comparison of dual-earner couples in three neighbourhoods of the Amsterdam metropolitan area highlights the significance of different life-style preferences – with respect to career advancement and approaches to child-rearing – for the kinds of neighbourhoods in which people choose to live. Simonsen's life-histories of women in a working-class neighbourhood of Copenhagen show that 'sense of place' and valued ties to enduring social networks foster continuity of residence, with implications for the way daily life is managed and experienced. García-Ramon and Cruz provide an example of the ways in which women's expectations about domestic work are tied to their thinking about place. The Andalusian women day-labourers complain because their husbands go to the bars rather than helping as much as they would like with the housework, but they also reject training their sons in housework because of social pressures that they believe are stronger in rural than urban areas.

Our studies also reveal the temporal dynamism of values and the importance of not casting place variations in stone. André points to the effects of revolution on attitudes towards the role of education in society, and the extent to which changing emphasis on education fostered women's entry into new areas of the Portuguese labour market; Vinay shows how women's expectations about combining work and family life in such a way as to maintain an acceptable quality of life are promoting changes in the organization of Italian society; and Sabaté-Martinez identifies emerging conceptions of more equitable gender roles, especially in relation to reproductive work among rural Spanish women. Indeed, the pace of change in gender values in southern Europe is a feature that recurs in our book (see also García-Ramon and Cruz, Duncan, and Kofman and Sales).

Finally, we draw attention to the existence of conflicting values within societies and at the household level This is a major issue in Chapter 9 by Eva Humbeck. While Germany favours policies of cultural uniformity rather than multiculturalism, immigrants, as exemplified by the Thai women, have more complicated sets of values – elements from the places of origin, values particular to their status as women seeking to change their own lives, and values they will acquire in their new country.

Further, their priorities and assessments do not coincide with those of their German husbands. Humbeck also shows how non-governmental (including women's) organizations express diverse values in their relations with the immigrant women. Clearly these issues of multiple and conflicting values will become more problematic as the EU attempts to harmonize social as well as economic policies. As Vaiou argues, 'Fortress Europe' already includes much diversity and 'in the process of integration, "others" are continuously constituted and deprived of rights that unified Europe is, in theory, promoting'.

DOES CONVERGENCE FOSTER GENDER EQUALITY?

The most conspicuous aim of the European integration process since its origins in 1957 has always been economic integration, and doubtless its main achievements have been economic as well, in spite of the proclamations of political and social goals. The Treaty of Maastricht confirmed this trend and its main concern was to establish a schedule for EU economies' 'convergence' towards a single currency. Of course, economic convergence has social implications, especially concerning labour market regulations and social welfare systems – both of paramount importance for women's work and daily lives. In effect, there is a European social policy, but its implementation has proved to be very difficult because of complexities arising from different institutional, political and cultural settings at the national as well as the regional and local levels.

Truly, since the 1957 Treaty of Rome, the EU has proclaimed its commitment to the principle of gender equality and has consistently developed policies (in the forms of specific Directives) and instruments (for instance, the European Parliament's Committee on Women's Rights and the Equal Opportunities Unit) to implement this principle in several spheres. But progress in that direction relies heavily on the member states' social and welfare regimes, which present strong differences rooted in different political evolutions and in the diversity of social and gender inequalities.

In spite of diverging interpretations of the principle of subsidiarity and the different conceptions of the implications of belonging to the EU, manifested, for example, in Britain's refusal of the 1989 European Social Chapter, a growing homogenization of social and welfare policies is likely to take place in the future. Several forces will push in this direction, such as the strengthening of European-wide policy in these fields or the inherent dynamics of economic integration in the Single Market (requiring homogenization of social charges and of social security claims). This foreseeable 'social convergence', even grounded on very general principles, is likely to have deeply different implications for women (as well as men) at the national as well as the regional and local levels. The effects will reflect the model towards which convergence is directed.

The increased presence of the Scandinavian countries may foster the commitment of the European Commission and the European Parliament to a more active social policy, especially in the field of gender equality. At the same time, the growing economic and political weight of Germany within the EU might have the opposite effect in fostering its own typical gender contract, that is, the 'housewife contract'.

What is clear is that the current trend in the EU seems to be dominated by an explicit neo-liberal agenda – part of a global scale tendency in the late twentieth century. On the one hand, the trend towards social benefits is being increasingly linked to participation in the (formal) labour force, thus further deepening the gender-based and other inequalities in the labour market. On the other hand, the state is gradually moving from provider to regulator of social services. Restrictions in the public provision of social services may be detrimental to women by reinforcing reliance on their traditional gender roles as caregivers for children and the elderly. As these processes play themselves out, women of the EU seem set to lose some of the benefits that have distinguished them from women in other parts of the world.

All in all, changes in the social and welfare regimes along these lines will disadvantage the low-paid, the part-time and the unemployed, as well as those outside the formal labour market, among which women are the most conspicuous, together with non-EU immigrant workers. The process of European integration is thus a process of construction of atypical 'others', many of them living inside what is being called 'Fortress Europe.' And women in the north as well as in the south, among nationals or immigrant minorities, are a large majority of these 'others'. We hope that one of the contributions of this collection of essays will be to highlight issues around which women can find common cause while remaining sensitive to the particularities of context.

NOTES

1 *The World's Women 1970–1990. Trends and Statistics* (United Nations Department for Economic and Social Affairs Statistical Office, 1991) is a valuable source of comparative data.
2 For a more detailed discussion of the complex politics and implications of subsidiarity see Scott *et al.* (1994) and Van Kersbergen and Verbeek (1994).
3 Jane Millar (1992) provides an extended discussion of the problems of executing a cross-national study of 'solo' women in the European Community.

REFERENCES

Arroyo, F. (1988) *El Reto de España: España en la C.E.E.*, Madrid: Sintesis.
Berthaud, C. (1991) *Le Marché Commun*, Paris: Masson.

Blacksell, M. and Williams, A. M. (eds) (1994) *The European Challenge: Geography and Development in the European Community*, Oxford: Oxford University Press.

Buckley, M. and Anderson, M. (eds) (1988) *Women, Equality and Europe*, Basingstoke/London: Macmillan.

Clout, H., Blacksell, M., King, R. and Pinder, D. (1989) *Western Europe: Geographical Perspectives*, Harlow: Longman.

Cockburn, C. (1992) 'The European forum of socialist feminists: talking on the volcano', *Women's Studies International Forum* 15: 53–6.

Cole, J. and Cole, F. (1993) *The Geography of the European Community*, London: Routledge.

Commission of the European Communities, Division IX/D-5 (1986) *Europe as Seen by Europeans – European Polling 1973–86*, Luxemburg: Office for Official Publications of the European Communities.

Connell, R. (1990) 'The state, gender and sexual politics: theory and appraisal', *Theory and Society* 19: 507–44.

Coopmans, M., Harrop, A. and Hermans-Huiskes, M. (1989) *The Social and Economic Situation of Older Women in Europe*, Brussels/Luxemburg: Commission of the European Communities.

Crompton, R., Hantrais, L. and Walters, P. (1990) 'Gender relations and employment', *British Journal of Sociology* 41, 3: 329–49.

Cunningham, S. (1992) 'The development of equal opportunity theory and practice in the European Community', *Policy and Politics* 20, 3: 177–89.

Dawson, A. H. (1993) *A Geography of European Integration*, London: Belhaven.

De Groote, J. (1992) 'European women's lobby', *Women's Studies International Forum* 15,1: 49–50.

Delacourt, M. L. and Zighera, J. A. (1992) *Women in the European Community: A Statistical Portrait*, Luxemburg: Office for Official Publications of the European Community.

Documents d'Anàlisi Geogràfica (1995) 26; special issue on 'Treball, occupació i vida diaria de les dones'.

Duchen, C. (1992) 'Understanding the European Community: a glossary of terms', *Women's Studies International Forum* 15: 17–20.

Elson, D. and Pearson, R. (eds) (1989) *Women's Employment and Multinationals in Europe*, Basingstoke/London: Macmillan.

Environment and Planning A (1994) 26, 8 and 26, 9; special issues on 'The diverse worlds of European patriarchy'.

Espace, Populations, Sociétés (1989) 1; special issue on 'Sexe et espace'.

European Urban and Regional Studies (1995) 2, 2.

Fagnani, J. and Chauviré, Y. (1989) 'La actividad profesional de las mujeres con hijos en la aglomeracion parisina (1975–1982)', *Documents d'Anàlisi Geogràfica* 15: 39–66.

Feminist Review (1991) 39 (Winter).

Garofoli, G. (ed.) (1992) *Endogenous Development and Southern Europe*, Aldershot: Avebury.

Geographica Helvetica (1994) 1; special issue on 'Geschlechterforschung in der Geographie'.

Geographische Rundschau (1995) 47, 4; special issue on 'Frauenbezogene Forschung'.

Hadjimichalis, C. and Sadler, D. (eds)(1995) *Europe at the Margins: New Mosaics of Inequality*, London: Wiley.

Hoskyns, C. (1986) 'Women, European law and transnational politics', *International Journal of the Sociology of Law* 14: 299–315.

—— (1988) ' "Give us equal pay and we'll open our own doors" – a study of the impact in the Federal Republic of Germany and the Republic of Ireland of the European Community's policy on women's rights', in M. Buckley and M. Anderson (eds) *Women, Equality and Europe*, Basingstoke/London: Macmillan.

—— (1991) 'The European women's lobby', *Feminist Review* 38: 67–70.

—— (1992) ' The European Community's policy on women in the context of 1992', *Women's Studies International Forum* 15: 21–8.

Iberian Studies (1991) 20, 1–2; special issue on 'Women, space and society in Spain and Portugal'.

Ilbery, B. W. (1990) *Western Europe: A Systematic Human Geography*, Oxford: Oxford University Press.

Inchiesta (1992) 22, 96; special issue on 'Donne del Sud'.

Jenson, J. (1985) 'Struggle for identity: the women's movement and the state in western Europe', in S. Bashevkin (ed.) *Women and Politics in Western Europe*, London: Frank Cass.

Kaplan, G. (1992) 'Contemporary feminist movements in western Europe: paradigms for change?', in B. Nelson, D. Roberts and W. Veit (eds) *The Idea of Europe: Problems of National and Transnational Identity*, New York/Oxford: Berg.

Katzenstein, M. F. and Mueller, C. M. (eds) (1987) *The Women's Movements of the United States and Western Europe*, Philadelphia: Temple University Press.

Kempeneers, M. and Lelièvre, E. (1993) 'Women's work in the EC: five career profiles', *European Journal of Population* 9: 77–92.

Lovenduski, J. (1986) *Women and European Politics: Feminism and Public Politics*, Brighton: Harvester/Wheatsheaf.

Masser, I., Sviden, O. and Wegener, M. (1992) *The Geography of Europe's Futures*, London: Belhaven.

Mazey, S. (1988) 'European Community action on behalf of women: the limits of legislation', *Journal of Common Market Studies* 27,1: 63–84.

Meehan, E. (1992) 'European Community policies on sex equality: a bibliographic essay', *Women's Studies International Forum*, 15: 57–64.

Meulders, D., Plasman, R. and Vander Stricht, V. (1993) *Position of Women in the Labour Market in the European Community*, Aldershot: Dartmouth.

Migliorini, F. and Pagliettini, G. (1993) *Città e Territorio nella Nuova Geografia Europea*, Milan: Etaslibri.

Millar, J. (1991) *The Socio-Economic Situation of Solo Women in Europe*, Brussels: Commission of the European Communities.

—— (1992) 'Cross-national research on women in the European Community: the case of solo women', *Women's Studies International Forum* 15: 77–84.

Minshull, G. N. (1990) *The New Europe into the 1990s*, London: Hodder & Stoughton.

Morokvasic, M. (1991) 'Fortress Europe and migrant women', *Feminist Review* 39: 69–84.

Noin, D. and Woods, R. (eds) (1993) *The Changing Population of Europe*, Oxford: Blackwell.

Pinder, D. (ed.) (1990) *Western Europe: Challenge and Change*, Oxford: Blackwell.

Power, L. (1991) 'The International Gay and Lesbian Association', *Feminist Review* 39: 187–8.

Prondzynski, I. (1989) 'The social situation and employment of migrant women in the European Community', *Policy and Politics* 14, 4: 347–54.

Scardigli, V. (1987) *L'Europe des Modes de Vie*, Paris: Centre Nationale de la Recherche Scientifique.

Scott, A., Peterson, J. and Millar, D. (1994) 'Subsidiarity: a "Europe of the regions" u the British Constitution', *Journal of Common Market Studies* 32: 47–67.

Solsona, M. and Treviño, R. (1995) 'Activitat, maternitat i paternitat a L'Europa comunitària', *Documents d'Anàlisi Geogràfica* 26: 191–207.

Springer, B. (1992) *The Social Dimensions of 1992: Europe Faces a New EC*, New York/Westport: Greenwood.

Todd, E. (1990) *L'Invention d'Europe*, Paris: Editions du Seuil.

United Nations Department for Economic and Social Affairs Statistical Office, Centre for Social Development and Humanitarian Affairs/United Nations Children's Fund/United Nations Population Fund/United Nations Development Fund for Women (1991) *The World's Women 1970–1990: Trends and Statistics*, New York: United Nations.

Vallance, E. and Davies, E. (1986) *Women of Europe: Women MEPs and Equality Policy*, Cambridge: Cambridge University Press.

Van Kersbergen, K. and Verbeek, B. (1994) 'The politics of subsidiarity in the European Union', *Journal of Common Market Studies* 32: 215–36.

Villanueva, M. (1994) *L'Europa Comunitaria: Les Transformacions del Territori en el Camí vers la Unitat*, Bellaterra (Barcelona): Servei de Publicacions de la Universitat Autonoma.

Williams, A. M. (1987) *The Western European Economy: A Geography of Post-war Development*, London: Routledge.

—— (1991) *The European Community: The Contradictions of Integration*, Oxford: Blackwell.

Windebank, J. (1992) 'Comparing women's employment patterns across the European Community: issues of method and interpretation', *Women's Studies International Forum* 15: 65–76.

Women's Studies International Forum (1992), 15, 1; special issue on 'A continent in transition: issues for women in Europe in the 1990s'.

Woolf, S. (1993) *Espaces et Familles dans l'Europe du Sud à l'Age Moderne*, Paris: Editions de la Maison des Sciences de l'Homme.

2

THE GEOGRAPHY OF GENDER AND WELFARE IN EUROPE

Eleonore Kofman and Rosemary Sales

INTRODUCTION

The continuing integration of European space is increasingly shaped by a neo-liberal agenda which fundamentally challenges the principles on which the postwar welfare states were built. As global economic competition undermines the ability of individual states to carry out independent economic policy, welfare policies are being subordinated to policies aimed at cost-cutting and maintaining profitability. Reliance on market solutions, and reductions in social expenditure, suggest that responsibility for welfare will increasingly fall on the family, and primarily on women (Cochrane 1993). The emerging model of social policy in the European Union is also likely to disadvantage women; by making benefits dependent on earnings, it compounds labour market inequalities.

These processes have led some to argue that the currently diversified gender and welfare regimes of the European Union will converge around a new model in which gender and other social inequalities will be markedly increased. At the same time, the end of the bi-polar division in Europe with the social and political transformation of Eastern Europe has brought the dismantling of the very different structures of welfare provision which had been developed in these states.

In this chapter, we argue that in spite of tendencies towards convergence, distinct welfare regimes remain, with differing implications for gender relations. The postwar years have seen substantial social and demographic changes which have tended to undermine the *male breadwinner model* on which the postwar welfare states were founded (Lewis 1993: 15). Women's involvement in the paid labour force has risen consistently and they have gained access to a widening range of occupations and professions. The size and shape of the family have been transformed, as fertility rates have declined and rising rates of divorce and births outside marriage increase the numbers of single-parent families.

Although the direction of these developments has been common throughout Europe, the form and pace of change have varied; substantial and sometimes even widening differences between states remain. The policy response of individual states to these new conditions has been different (Hantrais 1994). Just as the Keynesian approach to welfare was experienced differently in different states, so each is responding differently to the new 'austerity' (Cochrane 1993: 247). Even where, as in Britain, official policy has attempted to reverse these trends away from the 'traditional family', they have shown remarkable resistance.

Comparative analyses of welfare regimes have, with the exception of recent feminist writings, marginalized gender relations (Esping-Andersen 1990: but see, for example, Cochrane and Clarke 1993; Lewis 1993). There is a tendency, especially in analyses of industrialized societies, to take a narrow view of welfare, including only those elements which are quantifiable through some form of market exchange, and focusing primarily on policies aimed at supporting and regulating the labour market. Social policies in this definition fall into one of the following:

1 income maintenance for individuals unable to support themselves through the market or family – for example, unemployment benefit, poor relief, pensions
2 public provision of services – for example, education and health care, public housing, social work
3 regulation of working conditions – for example, the limitation of hours; exclusion of groups from particular types of work

All of these policies have explicit or implicit gendered implications. Welfare transfers produce different outcomes depending on the conditions of entitlement. The systems may be broadly earnings-related (with entitlement dependent on labour market participation), residual (strictly means-tested and intended only for those in greatest need), or universal (with entitlement based on citizenship). Earnings-related schemes tend to reinforce inequalities in the labour market and therefore disadvantage women. Residual systems are also likely to disadvantage women since means-testing generally takes the family income into account, assuming women's dependence. Universal benefits are the most equalizing.

In general, regimes which emphasize services rather than reliance on cash benefits are those where women's equality is more developed. Provision of services, particularly childcare, is essential for women's entry into the labour force, while the development of welfare services has provided the major expansion of employment opportunities for women. But it is the division of labour, and the structure of power and decision-making, as much as the quantity of spending on services, which determine their gendered effects.

Labour market regulations have important gendered effects. These may be explicit, as in the exclusion of women from certain occupations, or the provision of maternity leave; or implicit, such as the regulation of working hours, or the conditions of employment for part-time workers.

Both types of policy have major implications for women's ability to combine paid work with caring responsibilities.

Marshall's definition suggests a rigid distinction between social policy and the market:

> Social policy uses political power to supersede, supplement or modify operations of the economic system in order to achieve results which the economic system would not achieve on its own, and . . . in doing so it is guided by values other than those determined by open market forces.

> (Marshall 1964)

Welfare provision in practice, however, involves a mixture of state, market and voluntary agencies, often working in combination (German compulsory private health insurance schemes; British and Dutch voluntary agencies grant-aided by the state). While the dividing line between private and public is blurred, the neo-liberal agenda has increased the reliance on the private and voluntary sectors, although the mix varies in different European states. Following Chamberlayne's suggestive analysis of intermediate institutions, such as those connected with religious and political organizations and cultures of care, we need to broaden the way we think about welfare, from an exclusive concern with social protection and services.

According to Cochrane (1993: 5) a 'central feature of social policy in developed capitalist countries is to be found in the way it defines and constructs families as sources of informal welfare supply'. Conventional discussion of social policy tends to take the family, and the gendered division of labour within it, as given. Care undertaken by those employed by the national or local state, or through the market (in the public domain), is included as part of welfare, whilst care undertaken in the domestic (private) sphere is excluded from analysis. The family is a 'black box' that is outside the area of investigation (Brannen and Wilson 1987).

Feminist analysis has challenged this separation of the public and private spheres which is at the root of liberal thought (Pateman 1988). Male domination in the public sphere of employment and the state is based on the assumption of women's economic and sexual dependence on men within the family and the provision of unpaid care. By taking the family as a social construct to be analysed, rather than accepted as a natural fact, feminist analysis has been able to illuminate power relationships in both the private and public spheres.

In the development of welfare states, women have been central as clients, consumers and citizens, as providers of welfare, and as mediators on behalf of the family with welfare agencies. Women tend to be at the centre of a network of obligations, which includes not just children but parents and other relatives, neighbours and friends. We need to broaden our definition of welfare services to include support and care directed towards the physical, psychic and emotional well-being of people (Finch 1992). In the focus on quantifiable aspects of welfare, the less quantifiable elements of love and obligation, which these involve, are often ignored.

These obligations impact on women's citizenship status and entitlements. Caring responsibilities limit access to the labour market and thus women's entitlement to earnings-related welfare benefits (Waine and Jordan 1986). They also prevent women from participating in formal political structures to the same extent as men (Lister 1992). Few women occupy the higher strata of the welfare state, either as employees or elected politicians. While we do not assume a unified 'women's interest', the lack of women as decision-makers has important implications for the shape and distribution of welfare services. The development of refuges for women who have suffered domestic violence, for example, is unlikely to have occurred without campaigns and organization by women.

This chapter begins with a brief overview of the processes of European integration, which provides the context in which current welfare regimes are developing. Next we discuss one of one of the most influential comparative analyses of welfare in the work of Esping-Andersen (1990). While Esping-Andersen was not primarily concerned with gender, his characterization of 'welfare regimes' has been taken as the starting-point for much comparative work, including some which has used an explicitly feminist perspective. The subsequent section explores the typology developed by Lewis, in which regimes are characterized by how closely they approach the male breadwinner model. We argue that this approach, while providing useful insights, concentrates too narrowly on the labour market, ignoring other important dimensions to welfare. Drawing on Connell's (1990) analysis of 'gender regimes', we suggest how a theoretical framework that takes account of gendered state

Table 2.1 Women in national parliaments in the European Union

	Total	Women %
Denmark	179	33
Netherlands	150	31
Germany	662	20
Luxemburg	60	20
Spain	350	16
Belgium	222	14
Ireland	166	12
United Kingdom	651	9
Portugal	230	9
Italy	650	8
France	577	6
Greece	300	5

Source Parliamentary Labour Party 1994

structures, relations and interventions, might be developed. In doing this, we seek to incorporate the diversity of welfare access and outcomes for different groups, especially women in paid and unpaid work and immigrant and ethnic minority women (Cochrane and Clarke 1993; Kofman and Sales 1992). We conclude by returning to the question of convergence of European welfare regimes.

The introduction of privatization and dismantling of political structures in Eastern Europe is of vital importance to these debates; we believe, however, that these experiences are so distinct that we cannot do justice to them in this short chapter. We shall focus primarily on the states of the European Union and Scandinavia, though our discussion of Germany will take account of the bringing together of East and West Germany into a single welfare regime.

WELFARE POLICIES AND EUROPEAN INTEGRATION

With the crises in the Fordist modes of accumulation which underpinned the postwar consensus around Keynesian demand management and welfare policies in Europe until the 1970s, all EU states have embraced neo-liberal policies to varying degrees. Jessop (1993) has argued that the postwar *Keynesian Welfare State* (KWS) is being replaced by a new *Schumpeterian Workfare State* (SWS). While the KWS

> tried to extend the social rights of its citizenship, the SWS is concerned to provide welfare services that benefit business with the result that individual needs take second place.
>
> (Jessop 1993: 9)

Its distinctive features are 'a concern to promote innovation and structural competitiveness in the field of economic policy; and a concern to promote flexibility and competitiveness in the field of social policy' (Jessop 1993: 8). The Single European Market promotes this agenda, relying on cost-cutting rather than expansion to restore profitability (Grahl and Teague 1989). The Single European Act (1985) created an 'area in which the freedom of movement of goods, persons, services and capital is ensured'. Competition and flexibility are encouraged, while national measures to protect domestic industry are outlawed. The Maastricht agreement of 1993 marked a deepening of this strategy. The goal of a common currency necessitates harmonization of economic conditions between individual states. The convergence conditions are based on strict financial orthodoxy and require massive deflation for most economies, which has resulted in reduced and tighter social expenditure, for example, in Italy (Saraceno and Negri 1994).

The abandonment of the Keynesian Welfare State has exacerbated inequalities based on class, gender and ethnicity. The European Commission itself noted an increase in poverty in member states in the 1990s (Bennington and Taylor 1993: 126). But

[p]aradoxically, it is frequently those employed at the lowest levels of the welfare system who are most explicitly excluded from its benefits.

(Cochrane 1993: 16)

The postwar European welfare states were based on the shared assumption that women provided care generally in the private sphere of the home and were financially dependent on men. According to McDowell (1991), the old Fordist gender order in Anglo-American societies was based on a stable working class, the nuclear family supported by a male breadwinner and by women's domestic labour, and was underpinned by Keynesian economic and welfare policies. The male breadwinner has never been a general phenomenon in practice. Women have been forced to engage in paid labour either because of desertion or widowhood, if they are alone, or because their husband's earnings are insufficient to maintain the household. Migrant and ethnic minority women have generally been forced to engage in waged labour in either the formal or informal sectors while being excluded from many of the social benefits of welfare states (Williams 1989). But the ideology of the breadwinner has been pervasive and it underpinned the notion that women's work is less important than men's and legitimized unequal treatment (Barrett and McIntosh 1982).

Many states saw a substantial shift away from the male breadwinner model from the 1960s onwards, as women (including married women) entered paid employment in steadily increasing numbers. Neo-liberalism threatens a reversal of this trend: labour market deregulation exacerbates the marginal position of women (Bennington and Taylor 1993: 124), while the privatization of social reproduction and increased reliance on 'community' initiatives extends women's caring responsibilities. These caring responsibilities may make it more difficult for women to move around the European Union for work or training and thus to benefit from the Single Market.

While the burden of care of dependants is increasingly shifted on to family members (primarily women), these policies have also underpinned an expansion of women's part-time employment in some parts of Europe. This is facilitated by the restructuring of production processes and the development of 'flexible' work practices, which have been an integral element of a post-Fordist form of regulation. Analyses of this transition have generally neglected gender or relegated it to a passing note.

McDowell further demonstrates that new divisions have opened between and among men and women as a result of the restructuring of the labour market. While a minority of women have benefited from the expansion of professional and managerial positions, a larger number are pushed into casualized employment with low pay and limited benefits, and unemployment among men has risen throughout the European Union. Yet, the deepening of unemployment in many European states has recently led to new strategies to push women back into the home (see Fagnani 1994 for France; Ostner 1994 and Wilson 1993 for Germany).

As the completion of the Single European Market extends the free movement of capital, commodities and labour within the European Union, development of a 'Social Europe' has remained limited and fragmented (Bailey 1992; Liebfried 1993). The Social Charter of Workers Rights and the Social Protocol of the Maastricht Treaty proposed certain minimum standards aimed at mitigating the worst ravages of competition unleashed by the Single Market (social dumping). But these involved only limited measures, many of which are existing practice in some member states: indeed, some have better provision, prompting fears of 'levelling down' as well as 'levelling up'. These measures would nevertheless improve conditions in some states and the framework has acted as a catalyst for legislation supporting equal treatment of men and women, for example, in Ireland.

The EU's approach to social policy has mainly been confined to attempts to define common goals rather than the creation of common institutions. It is primarily regulatory rather than involved in direct welfare provision, although it can call upon a battery of financial policies with a social dimension through its structural funds directed towards poorer regions. The objectives of the Social Protocol focus on the promotion of employment, improvement of living and working conditions and adequate social protection through financial benefits for those unable to work.

The focus on *workers' rights* draws a firm line between the public and private spheres. It leaves untouched groups outside the formal labour process, such as the elderly, or the 15 million unemployed in the EU (Doogan 1992). It also excludes intervention in questions of sexuality and family policy. Issues of family and gender relations are central to the construction of national identity (Yuval Davis and Anthias 1989) and there has been little attempt to harmonize policy in these areas.[1] These matters are to be resolved through the principle of subsidiarity, leaving them largely in the hands of individual states (Cunningham 1992). Progress has therefore been slow, and obstructed by the British government which has maintained its 'opt-out' from the Social Chapter.

The process of integration stalled to some extent in the 1990s, as public opposition to unification, expressed, for example, in referenda on the Maastricht Treaty,[2] grew. The repercussions of German unification also slowed down the momentum towards centralization; Germany's economic dominance has allowed it to play a leading role in the integration process, but the economic and social strains of the unification process have weakened its position. While the European Union extends outwards to embrace Austria and Scandinavian countries such as Sweden and Finland, the process of deepening economic and political union has reached stalemate with the crisis in the Exchange Rate Mechanism.

The one area where centralization has intensified is immigration policy. Intergovernmental institutional arrangements have been established to harmonize policy outside the scrutiny of the European Parliament and Commission (Webber 1993). The term 'Fortress Europe' has been widely used to describe these processes of exclusion which have been primarily aimed at people of non-European origin (Bunyan 1993; Kofman and Sales 1992).

Any movement towards a European-wide conception of social citizenship will make it more important 'to define who is "in"' (Liebfried 1993: 12). European integration is producing two sets of citizenship rights as a new Europeanized framework for nationality is developed. EU nationals have freedom of movement within the EU, access to employment and welfare rights; non-nationals have limited freedom of movement and no absolute right to employment or welfare. As Taylor (1989: 27) argues, 'the integrative advantages of welfare policies for some sections have been inextricably linked to the denial of citizenship for others'. Recent changes in asylum laws across the EU have reduced access to welfare for refugees – for example, in public housing – and have tightened conditions for claiming benefits (see, for example, Refugee Council 1993).

While curbs on permanent settlement have been strengthened, the increase in flexible work practices, including casualized labour and home-working, has maintained a demand for labour whose status is insecure and open to super-exploitation. Many European states have seen an influx of temporary workers; for example, the overwhelmingly female domestic workers from Asia.

The social policy of the EU remains contested, as the European Commission and the European Parliament have tended to press for a more radical agenda than the Council of Ministers. While national parliaments veered to the right, after the European elections in June 1994, Socialists became the largest group in the European Parliament. With the development of an Equal Opportunities Unit and European Parliament Women's Committee, proposals are being developed which would go beyond promoting formal equality towards providing conditions for substantive equality through, for example, the funding of childcare provision in member states. In July 1994 the Commission brought forward proposals to end Britain's opt-out from the Social Chapter and to include for the first time anti-racist measures.

WELFARE STATE REGIMES

Esping-Andersen's comparative study of welfare states (Esping-Andersen 1990) has been highly influential: his typology of *welfare state regimes* has been taken as the starting-point for much subsequent analysis of European welfare (Cochrane and Clarke 1993; Johnston 1993; Lewis 1993).

His approach to the development of welfare states is based in political economy. He focuses not merely on the narrowly defined area of state social spending but on 'the state's larger role in managing and organizing the economy' (Esping-Andersen 1990: 2) and how its activities interlock with market and family.

Esping-Andersen suggests that in the postwar period advanced capitalist countries saw the emergence of a number of distinct *welfare state regimes* which he labels liberal, conservative and social democratic. Decommodification, social stratification and employment are keys to a welfare state's identity (Esping-Andersen 1990: 2). The critical aspect of the benefit system is its capacity for 'decommodification', or 'the degree to which individuals, or families, can uphold a socially acceptable standard of living independently of market participation' (Esping-Andersen 1990: 37).

The *liberal regime* is typified by English-speaking former colonies of settlement such as the USA and Australia. Welfare provision is strictly means-tested and adequate to maintain only very modest living standards. Thus the middle classes rely heavily on the market to supplement state provision. Although 'in the liberal ideal, concerns of gender matter less than the sanctity of the market' (Esping-Andersen 1990: 28)

> the liberal dogma is forced to seek recourse in pre-capitalist institutions of social aid, such as the family, the church and the community, and in doing so, it contradicts itself, because these institutions cannot play the game of the market if they are saddled with social responsibilities.
>
> (Esping-Andersen 1990: 42)

The *conservative regime* developed in corporatist states, such as Bismarck's Germany, France, Austria and Italy. Benefits are high but largely earnings-related with little role for private insurance schemes, and redistributive effects are negligible.

The conservative nature of these regimes is also expressed in the promotion of 'family values', and a strict differential between men as breadwinners and women as wives and mothers. Married women are discouraged from paid work. The care of dependants is presumed to be a private family matter, falling mainly on women. The welfare state is governed by the principle of subsidiarity, intervening only when the family's capacity to care for its members is exhausted. The strong influence of the Church both supports family ideology and itself provides an important source of welfare.

The *social democratic regime* is exemplified in Scandinavia, where the benefit system is based on universalism and a high degree of decommodification. Support for high taxes is maintained by benefits sufficient to meet middle-class aspirations, leaving little role for private provision. There is a high degree of socialization of family responsibilities with well-developed childcare facilities. These regimes depend on full employment both of women and men to maintain political support for high taxes and to finance benefits.

A number of critiques have been levelled at Esping-Andersen's typology. While some have addressed the choice of criteria he has used, and have suggested alternative 'regime types' (Liebfried 1991), our concern here is with those which highlight the inadequate theorization of gender in his characterization of welfare regimes.

Esping-Andersen focuses on formal employment and the welfare entitlements that it yields. His characterization of the different regimes has clear gender implications, but because gender is not central to his theorization women enter the analysis only when they are in paid work (Lewis 1993). The family and social reproduction are important elements in distinguishing his regime types, but it is their implications for access to paid work which interest him. He does not explore the gendered division of labour, or power relations within either the public or private spheres.

Welfare is seen in terms of the transfer payments (social security, unemployment benefits, pensions) which make up his index of decommodification. This focus on commodification obscures the extent of welfare provided through unpaid domestic labour which is a necessary basis for the commodification of labour, and the extent to which the form of decommodification is itself gendered. Unpaid work, as British research reveals, is substantial, estimated at £24 billion in wage equivalent in 1990 (Taylor-Gooby 1991: 100–1). Langan and Ostner (1991: 131) point out that

> [m]en are commodified . . . by the work done by women in the family. Women on the other hand, are decommodified by their position in the family. Thus men and women are 'gendered commodities' with different experiences of the labour market resulting from their different relationship to family life. . . . [Women] protect the vulnerable in a market economy, that is those who cannot become fully commodified but this in turn makes them vulnerable.

THE GEOGRAPHY OF WELFARE REGIMES

Esping-Andersen's coverage is directed towards distinctive welfare regimes derived from the experiences of specific core states and their balance of class forces at key moments in the evolution of state- and nation-building. Thus for the liberal his exemplar is the USA; for the conservative corporatist Germany; and for the social democratic Sweden.

Liebfried (1991) develops a typology broadly corresponding to Esping-Andersen's, but introduces a fourth regime type, which he calls a *Latin Rim* to characterize states of the southern periphery (Spain, Portugal, Greece and, to a lesser extent, Italy). Ireland has also sometimes been placed in this category. Liebfried classifies these regimes as *fundamental*, that is, with relatively low provision by the state. These are similar to liberal regimes in stressing residualization, but they can call upon older religious traditions of welfare as well as the family. They retain a strong agricultural basis, a rudimentary subsistence economy and a thriving black economy.

This characterization treats the European periphery in a time-warp, perpetuating a simplistic north–south model that ignores the development and expansion of welfare systems, which took place in the 1970s and 1980s. It is particularly difficult to sustain in relation to Italy, whose welfare system expanded from the late 1960s, and shifted from what had been a minimalist system to a universal model of social democratic type, especially in the north-west and Emiglia-Romagna (del Re 1993). The national health service set up in 1978 is free of charge to all citizens, while social security was extended to cover illness, accident, maternity leave, old age, invalidity and unemployment (Bimbi 1993: 152). The system is characterized by clientism (Chamberlayne 1992a) with the boundary between political and welfare systems blurred (Paci, cited in Chamberlayne 1992b). The provision of services is therefore highly uneven, being much better developed in the richer areas of the north and centre than in the south (Bimbi 1993: 165).

Ireland too has developed a wide range of measures since the 1970s within a conservative corporatist model in which the Catholic Church retains a strong influence (McLaughlin 1993). In Greece, Portugal and Spain, the transition from authoritarian rule in the 1970s and the continuing reliance on varying insurance modes suggest that these states are moving from fundamental towards corporatist regimes.

Some writers have claimed that there is a distinctively Catholic style of welfare regime. As we have already suggested, this is too simplistic an analysis of the regimes in the southern periphery and Ireland. Catholicism has left its mark in terms of ideas and practices far more widely than in these areas: for example, the concept of subsidiarity in Germany, now used in a modified form by the European Union, derives directly from Catholic social ideology (Chamberlayne 1992b; Manning 1993). Chamberlayne (1992) points out that in all states in which Catholicism plays or has played a significant role, intermediate organizations are a strong feature of welfare systems.

The geographical pattern of church involvement within states may be quite varied: for example, the Catholic Church and the Communist Party dominate in different localities in Italy; in Germany and the Netherlands there is a plurality of religious organizations; while in Ireland the Church's role in welfare (particularly education) is formalized within the state, and has actively promoted a traditional breadwinner model (Mahon 1994). Jenson (1991) shows how in the early postwar years the Church was influential in French family policy, but was later displaced by other political discourses more open to neutrality on the shape of the family.

States whose systems are hybrids, with elements of more than one type of regime, are poorly conceptualized in Esping-Andersen's schema. The British system, for example, while increasingly liberal, retains universal and social democratic aspects. Esping-Andersen classifies France alongside Germany and Italy as conservative and familial. But whereas the benefit system encourages women to combine paid work and childcare in France, the German system discourages it. Having classified France as conservative, according to his key indicators, he ignores the fact that its childcare provision is comparable with Scandinavia's: 'Bismarck plus childcare' (Grignon 1993: 51).

Another area where the gender dimension breaks down his categories is Scandinavia, which he treats as a relatively undifferentiated regime-type. Leira (1993) argues that Norway treats women primarily as wives and mothers, and that its childcare provision is more similar to the British model than to the social democratic. Esping-Andersen also fails to acknowledge the importance of the gendered division of labour in unpaid caring work. Socialized childcare does not exhaust the responsibility for caring with even the most advanced provision. Although labour market policies may aim towards gender neutrality, the sexual division of unpaid work means they have different consequences for women and men (Siim 1993: 34). In Denmark, in particular, a large proportion of women organize their paid work around childcare, and are in the labour force part-time. Part-time workers in Denmark have access to employment benefits comparable on a *pro rata* basis to full-time workers, which places them in a better position than part-time workers in, for example, Britain.[3] Nevertheless, the difference in working practices is a major element of gender inequality.

In times of welfare austerity and increased reliance on mixed welfare provision, including community initiatives, geographical differences within states are likely to become more significant. In many European states there has been a substantial shift towards a more decentralized model. Local and regional policies and institutions have become increasingly important in modifying access to welfare services for different groups of women and for non-citizens. These intersect with geographical variations in family structures and traditions of care.

Italy, for example, has decentralized its welfare provision since the 1970s. In the more economically developed areas of northern and central Italy, there is a well-developed system of social services in which women have built up a solid basis of welfare based on citizenship rights rather than social assistance (Bimbi 1993: 163). The territorial divide coinciding with a social dualism has become clearer since the 1980s (Bimbi 1993: 165–6), with birth rates falling faster in the north than in the south. Geographical differences are also significant in Germany with its dual welfare system based, on the one hand, on social security posited on a model of regular and continuous employment, and, on the other, on social assistance for those completely or partially outside employment.

The conclusion of Sackmann and Haussermann (1994) from their study of Germany would be equally applicable to Italy:

> regions should not be seen as following one predetermined path of development, where there is just one way of integrating women into the labour process. . . . Rather, they represent 'regional societies' characterized by contrasting features of the work spheres of gainful employment and household work, and how they are integrated or separated.

Even in Sweden, which is much less regionally divided, the distribution and composition of services such as salaried child-minding and publicly funded municipal day-care is quite uneven (Ginsburg 1993: 195–6).

The British welfare state has seen both a fragmentation of service provision and a centralization of resources and decision-making. Recent legislation allowing schools and hospitals to 'opt out' of local authority control has reduced the role of locally elected bodies as providers, while giving more power both to central government and to the private sector. This has made it more difficult for local authorities to develop their own initiatives or to make substantial improvements to existing provision.

GENDER AND WELFARE REGIMES

A number of writers have taken the framework developed by Esping-Andersen or Liebfried as a starting-point for a gendered comparative analysis of welfare regimes.

Langan and Ostner (1991) base their feminist framework on Liebfried's typology. They make an important critique of Esping-Andersen's ungendered theorization, and their analysis of the welfare regimes of Scandinavia, Germany and Britain provides a useful amplification to our understanding of the gendered nature of these regimes. They characterize the social democratic regime as a 'universalisation of a female social service economy'; the German 'Bismarckian model' as a 'gendered status maintenance model'; and the Anglo-Saxon regime, which they argue applies increasingly to Britain, as a dualistic model. This assumes full equality in the market place but at the same time is based on a model of the family in which 'wives and others are expected to be absent from public life as non-working dependants' (Langan and Ostner 1991: 139).

Their analysis, however, is limited by an uncritical adherence to Liebfried's typology, particularly in relation to the Latin Rim. Discussion of the latter is sketchy, showing little acquaintance with developments since the 1970s, and, by concentrating on the 'ideal-types' of this typology, they continue to treat regime clusters as homogeneous. For example, they do not mention France, which these typologies assign to a corporatist or status maintenance model, although its gender regime is markedly different from that of Germany.

Taylor-Gooby (1991) emphasizes the relationship between paid and unpaid work and the private and public spheres. Using Esping-Andersen's typology he applies a corresponding gendered division operating along the private/public axis. This in turn, he argues, generates specific forms of gender struggles.

In social democratic regimes, especially Denmark and Sweden, socialization of care has taken place through the expansion of the welfare state, which has extended women's entry into formal employment and necessitated dual-income families to maintain high taxation and benefits. This has created a high level of horizontal and vertical gender segregation with women concentrated in the state sector. It is therefore likely that future struggles will occur in the sphere of paid employment and for the preservation of welfare citizenship. Conservative corporate regimes tend to define care as the preserve of women located in the informal and private spheres, often organized by religious institutions. Gender struggles are most likely to centre on the extension of welfare citizenship, and on the recognition of care in the system of occupationally stratified rights. In the liberal regime, the increasing commodification of care-work means that access to caring services (such as childcare) is increasingly dependent on ability to pay. Inequalities in the labour market are compounded by inequalities in this sphere, exacerbating class divisions. It will be working-class women in particular, who cannot afford these services, who will struggle for the extension of welfare citizenship.

We would question whether one can derive the nature of gender struggles without reference to existing women's movements and the complex interplay between class, gender, sexuality and ethnicity. In liberal regimes, for example, the loss of income and opportunities for middle-class women caused by the reshaping of welfare provision, as much as the nature of women's organizations, may push middle-class women into the forefront of claims against the state and an extension of welfare citizenship. For many ethnic minority and immigrant women, issues of nationality and the basic rights of citizenship are the major priority (Bhabba and Shutter 1994).

This schema does not address important areas of gender struggle concerning rights over the body, reproduction and sexuality, which cannot be reduced to normative welfare claims. Campaigns over the right to abortion have been the major focus of gender struggle in Germany and Ireland in recent years. These struggles too have different implications for different groups of women. Black and ethnic minority citizens experience different relationships to welfare services, particularly in access to control of fertility. White women are generally encouraged to reproduce and black women discouraged (Anthias and Yuval Davis 1992). The former often face obstacles to abortion and sterilization, while the latter are much more likely to be offered both, as well as unsafe methods of contraception such as depo provera (Klug 1989).

Lewis (1992), however, bases her gendered typology on the extent to which the 'male breadwinner' model has been retained. This issue focuses directly on the gendered form of decommodification, and on the links between unpaid and paid work. In the breadwinner model, men gained a 'family wage' and direct access to benefits, while women gained welfare entitlements by virtue of their dependent status within the family as wives and mothers.

THE MALE BREADWINNER MODEL

Lewis divides states into strong, modified and weak 'male breadwinner' models. As examples of *strong male breadwinner* states she selects Ireland (conservative corporate) and Britain; we would also add Germany. In Ireland the state and the Church together ensured that women's primary role was within the family as mothers, carers and dependants. This role was enshrined in the 1937 Constitution. The marriage ban against women working in the civil service was lifted only in 1977 and until 1984 married women received lower rates of benefit, for shorter periods, and were not eligible for unemployment assistance. Lack of welfare and employment rights has been a major factor in the high rate of female emigration, which since the end of the nineteenth century has consistently outnumbered the male rate (Rossiter 1991).

Ireland still has the lowest percentage of active women in the age group 25 to 49 years, with less than 30 per cent of mothers in employment (Pillinger 1992). Crèche facilities are scarce and women are expected to look after elderly relatives. Reform of social security in the mid-1980s was prompted in large part by EC law on equal treatment of men and women in social policy (McLaughlin 1993). This, together with the activities of women's groups, has begun to make inroads into the rigid male breadwinner model. The participation of married women increased from 8 per cent in 1971 to 27 per cent in 1991. The increase has been greatest among urban middle-class women. Mahon (1994), following Walby's analysis (Walby 1990), describes this development as the move from private to public patriarchy. In 1991, for the first time the proportion of women engaged in 'home duties' (private patriarchy) dropped below 50 per cent (Mahon 1994). Ireland nevertheless retains one of the strongest breadwinner models in Europe.

Britain has historically followed a strong breadwinner model. To Beveridge, whose report laid the basis of the welfare state, the concept of full employment applied only to the masculine half of the population. Women as wives and mothers were dependent on the male wage. Married women were offered the possibility of paying lower contributions and consequently collecting less benefit. The invalid care allowance (for caring for a sick relative) has been denied to married women on the grounds that such care-giving is part of their 'normal' duties (Lewis 1992: 164).

Although Britain and France display similar levels of female labour force participation, the difference in composition is striking. The expansion in married women's participation in Britain can be accounted for by the increase in part-time work, particularly short part-time (less than sixteen hours per week), which British employment law makes especially precarious. In France, on the other hand, only 23.5 per cent of women workers worked part-time in 1989: many of these were either younger women looking for their first job or older women. The conditions of part-time workers are also markedly different: in France each hour of employment is counted as 'social' time (i.e. it involves pension and other contributions); while in Britain these rights do not exist where less than eight hours per week are worked.

In its embrace of 'Victorian values' the Conservative government has sought to address the changing reality of family structures by exhortations to the family to shoulder the burden of care of children, the disabled and the elderly. British childcare provision is among the lowest in Europe, with 2 per cent of children under 3, and 35–40 per cent of those from 3 to 5 years (the age of compulsory schooling) in publicly funded day-care. Children are viewed as the 'private property' of individual families (Edwards and McKie 1994: 58), which are forced to make their own arrangements for childcare.

Britain has the second highest percentage of lone parents (15 per cent) in the European Union and the highest divorce rate at 2.9 per 1,000. Single mothers are treated as mothers rather than breadwinners. The lack of childcare arrangements and the disincentive effect of the security system have meant that Britain is the only European state where lone mothers have lower labour force participation than married women.

West Germany (the FGR) followed an explicitly male breadwinner model in which marriage provides women with social protection and a basic income (Pfau-Effinger 1993: 391). While the more overt forms of patriarchal power have disappeared – until 1977, for example, a man could prevent his wife from taking on paid work (Ostner 1993: 98) – the benefit system continues to discriminate against working married women. Childcare provision is poor, while a major obstacle for working mothers is that children normally spend only half the day at school, with no lunch provided (Pfau-Effinger 1993: 392). Marriage is privileged both ideologically and within the benefit system. In contrast to the low labour force participation of married mothers, 76 per cent of single mothers are in paid work, much of it poorly paid (Ostner 1993: 105).

The gender regime of the former GDR was very different, based on almost universal labour force participation, including mothers. This was made possible by a system of heavily subsidized workplace and district kindergartens (nurseries). The policy measures aimed at reconciling paid work with family responsibilities were, however, addressed exclusively to women. This tended to reinforce traditional gender divisions of labour within the family (Quack and Maier 1994: 1259). Although women had entered non-traditional sectors, gender segregation remained strong, with the feminization of both public and private childcare (Ostner

1993: 107). The system did not discriminate in favour of marriage, making single motherhood increasingly popular.

With the victory of the conservative coalition in the 1990s East German elections, the way was cleared for the incorporation of East Germany into the Federal Republic, the phasing out of the old regime's more liberal family law and the dismantling of their social provision. This process coincided with the collapse of the East German economy, which has brought rising unemployment in which women's jobs have been particularly vulnerable (Wilson 1993: 165).

Faced with reduced employment opportunities, women have fewer options available than men, since their mobility is more constrained by domestic responsibilities. Almost 2 million people have left East Germany since 1989, mainly for the West: many more live in East Germany but commute to West Germany for work, with a substantial number travelling such long distances that they are unable to return home on a daily basis. The majority of both migrants and commuters are men (Quack and Maier 1994: 1272).

In spite of these developments, the adoption of a male breadwinner model has been resisted, with women demonstrating continued attachment to labour force participation. The level of women's economic activity has fallen as a result of high unemployment rather than an ideological acceptance of economic dependence (Quack and Maier 1994). This resistance has so far been mainly individual and private. Quack and Maier fear that as women disappear from paid work with increased unemployment, they are less likely to be able to develop public and collective forms of protest (Quack and Maier 1994: 1273), particularly since there is little tradition of specifically gender struggles on which to draw. Nevertheless, the refusal of women in the East to embrace a familial model may serve to undermine the breadwinner ideology of the West, and provide the basis for common struggles with West German women.

In France's *modified male breadwinner* system, family policy has been dominant within the social security system, while the child is the centre of family policy. From the postwar years family policy has been the 'cornerstone of the modernisation of social policy' (Jenson 1991) and a central element in the constitution of Fordism. Its primary aim has been to compensate parents for the costs of children, the priority being a horizontal distribution of resources towards families with children rather than towards lower-income families. Discussions at the end of the 1960s on how to use family policy to tackle social inequalities led to the state recognizing the existence of lone parents and the reality of working mothers.

France has one of the highest levels in Europe of women in full-time employment, especially of mothers with children under 10 years. Its publicly provided childcare is also among the best in Europe. State policy has increasingly sought to make child-bearing compatible with employment. Irrespective of the number of children, a higher percentage of French women continue to work after childbirth than German women. Recent proposals for legislative change would, however, shift this emphasis: for example, parental leave (*allocation parentale d'education*), currently paid only to those with three children, may now be paid to those with two children in order to encourage more women to leave the workforce at a time of unemployment (Fagnani 1994). Similarly the introduction in 1992 of *Aide aux familles pour l'emploi d'une assistante maternelle agréée*, or cash to employ registered childminders, can be seen as a strategy to turn unemployed women into day-care providers and reduce expenditure on day-nurseries.

The *weak male breadwinner* model applies to Sweden, as well as to Denmark and Finland. In Sweden the social democratic vision of women in the 1940s was in terms of two sequential roles: mothers and workers. Welfare spending until the 1960s was below the OECD average and it was not until the 1970s, after social pressures for greater class and gender equality, that social expenditure increased dramatically. Since the mid-1970s women have been treated as workers and compensated for their unpaid work as mothers at market rates. Separate taxation was introduced in 1971 and public day-care increased massively in the 1970s. A scheme of parental leave to compensate for loss of earnings, instead of flat rate maternity benefits, was also implemented in 1974. Women now have one of the highest labour force participation rates and by far the highest hourly rates of OECD states (90 per cent of men's).

The model has not, however, overcome gender divisions but has entrenched them in new forms. Entitlements such as parental leave continue to be largely taken by women, which has consolidated the gendered division of labour in both private and public spheres. Sweden has one of the most horizontally and vertically segregated employment patterns in Europe.

CARE OF THE ELDERLY

The incorporation of the breadwinner model adds a necessary dimension to our thinking about welfare regimes. It is based, however, on a relatively limited conception of welfare with its focus on the nature of women's access to the formal labour market. Within this framework, discussion tends to concentrate on childcare as the dominant responsibility. Public intervention has touched to an even lesser extent care of the elderly and disabled. While childcare facilities and benefits are generous in Belgium and France, for example, neither residential care nor home help and nursing for the elderly have been given much prominence (Jamieson 1991) and the elderly are largely to be supported by the family.

The European Commission forecasts that in the next century, all EU member states will face rising dependency ratios, with increases in the number of pensioners, particularly of people aged over 80 (*Guardian*

5.1.95). As Bennington and Taylor (1993: 126) argue, one of the 'crucial contradictions facing the welfare state in relation to care for elderly people is the dual role played by women within society and the economy'. Members of today's working generation face longer full-time education for their children and an increasing number of very old relatives. However, the impact on family carers, largely women, is not simple: some give up paid work, while others carry a double or even triple burden.

We can make the distinction between obligatory and non-obligatory familial support for the elderly in welfare policy (Ostner 1994: 38–9). Denmark has the fewest statutory family obligations with the most cases of elderly in need of care located in publicly funded institutions. At the other extreme, the constitution of many European states (those categorized by Esping-Andersen as 'conservative') includes an obligation of parents and children to care for each other. While family policy in France and Germany has been markedly different, this reflects a greater degree of familial solidarity in these states than in the more atomized societies of either Britain or the social democratic regimes of Scandinavia. These obligations extend to the care of adult children: in France, for example, young people are not eligible for benefits until 25, while the expectation that children set up home independently before marriage (and live away from home while studying at university) is a largely British and Danish phenomenon.

In several countries, especially those in the Latin Rim, although the family may be recognized as a major provider of welfare, changes in family structure and family care have simply not been discussed by policy-makers (Dell'Orto and Taccani 1992 on Italy; Mestheneos and Triantafillou 1992 on Greece).[4] In Italy, the level of fertility has reached a record low of 1.3 children per woman whilst the proportion of elderly will be the highest in the European Union by the next century. Elderly people in need of care are dependent on the family, with virtually no public support, or on the Church, which provides residential care for those who can afford it.

For the elderly, the key distinction determining their right to care is whether their needs are considered medical, which qualifies for health insurance, or social, which does not. Home help and other domiciliary services are generally very poorly developed (Jamieson 1991). In Britain, where until recently elderly people had universal access to a range of medical and social services, the distinction is becoming crucial. The former remains free at the point of delivery while the latter may be severely means-tested by cash starved local authorities (*Guardian* 1994).

A crucial aspect of caring services is the relationship between the caring and the cared for. Feminist research has made the distinction between caring *about* and caring *for* a person to highlight the social relations of caring, and the complex mixture of labour and love it involves (Finch 1992). The tendency to focus on the carer while ignoring the needs, desires and rights of the person cared for has been brought out most forcefully in relation to people with disabilities (McLaughlin and Glendinning 1994). Feminists have also highlighted the blurring of the distinction between the individual private arena and the public arena in care (Leira 1992).

Relationships between carer and cared for, and paid and unpaid care, are structured by class and ethnicity. For example, black women in Britain have been concentrated in poorly paid public caring employ-ment (Williams 1989) and are increasingly employed in domestic service (Graham 1991). Professional women increasingly use the services of low-paid informal child-minders to allow them to maintain full-time employ-ment. In states with poor public care facilities immigrant women provide much of the domestic care (Andall 1992). Instances of this are increasing sharply in urban areas of Mediterranean states, where women have entered formal paid employment in greater numbers. A study for the International Labour Office (ILO) estimated that about half the foreign domestic workers in Spain (overwhelmingly female) are illegal, and are therefore extremely vulnerable to exploitation by employers (Collectivo IOE 1991).

The voluntary sector, often with a heavy religious presence, makes a substantial contribution to the care of the elderly and people with disabilities in all but the social democratic regimes (Chamberlayne 1992a). These elements are increasingly likely to figure in any plural welfare system.

A major problem in the classifications discussed so far is the absence of discussion of a political arena in which a variety of actors come into conflict and engage in struggles around specific issues (but see Jenson 1991 for France; Bimbi 1993 and del Re 1993 for Italy; Conroy Jackson 1993 for Ireland). Esping-Andersen, for example, bases his typology on class alliances forged at the time of the formation of the welfare system, but ignores this aspect in his discussion of contemporary developments. Inevitably, the weakness in the theorization of the political dimension underplays the role of women's movements and the alliances they have created at national, regional and local levels. In some of the least well developed welfare systems, where traditional family structures have remained strong (for example, in Italy), women's movements have succeeded in challenging gender practices in social reproduction and welfare provision.

GENDER REGIMES

So far we have discussed studies of the relationship between paid and unpaid work and care through the nexus of the state, intermediate institutions, the market and the family. We need to consider the overall relationship of women to the welfare state, in order to explore the role of the state in the formation and maintenance of the *gender regime* (Connell 1990; Franzway *et al.* 1989). As Connell (1990: 530) argues:

> The state is constituted within gender relations as a central institution of gendered power . . . it has a major stake in gender politics which is not fixed.

According to Connell, each empirical state has a definable 'gender regime' that is the precipitate of social struggles and is linked to – though not a simple reflection of – 'the wider gender order of society' (Connell 1990: 523). Three main structures contribute to this gender regime: first, a *gendered division of labour* within the state apparatus; second, a *gendered structure of power* in the state apparatus; and third, a *gendered structure of cathexis*, or emotional attachments.

This analysis raises wider issues about the role of the state in establishing and reproducing gender relations. An exploration of Connell's three structures suggests fruitful areas for comparative work. His concentration on state structures, however, is too narrow; we need to include state policies and their outcomes, and the interplay between the state and other institutions of socialization and welfare, and social and political movements. We would suggest the addition of these elements into Connell's framework to encompass the positioning of the state in relation to economic, social and political forces in the construction of the wider gender order of society.

Connell argues that the *gendered division of labour* within the state apparatus concentrates women in caring roles, while the elite and the coercive apparatus are dominated by men. We have seen that even in the most egalitarian models, caring work is dominated by women both in state employment and in the family. Individual welfare agencies, however, display both horizontal and vertical segregation, with men tending to occupy the top echelons of even such female-dominated professions as teaching and social work (see, for example, Inner London Education Authority 1986). The expansion of state employment for women in the postwar period has been the main vehicle by which private patriarchy has been transformed into public patriarchy (Walby 1990; Siim 1993).

The gendered division of labour itself embodies inequality: control of the coercive apparatus – including the judicial system for example, gives greater power than domination of caring services, and is itself central to the maintenance of gendered power. The state plays a major role in regulating sexual politics and in determining the conditions under which the deployment of sexual violence (whether against women or gay people) may be used.

Feminists have challenged the conventional view that the state has the monopoly of legitimate violence, pointing to the family as an arena of legitimized male violence. The control which marriage gave to men over women's bodies was a central element in the fraternal contract (Pateman 1988): the role of the judiciary in condoning male violence and rape in marriage has been crucial in cementing that control. While most European countries have recognized rape in marriage as a crime, they display marked differences in the way in which the judicial system deals with these issues in practice.

The right to control over fertility has been a major area of gender struggle. Although only Ireland completely outlaws abortion, in all but the Scandinavian countries women's right to abortion is conditional on approval from the male-dominated medical profession. Apart from legal restrictions, women's ability to enjoy this right in practice depends on the priority given to this area in the provision of health and social services, and in all states, these services are geographically uneven.

While Connell is concerned with the division of labour within the state structures themselves, state policies play a wider role in the regulation of the division of labour in both private and public spheres. These policies have implications for the extent of attachment to the breadwinner model.

The second element of Connell's schema, the *gendered structure of power*, encompasses both bureaucracy and personal links, the informal organization of resources and contacts, and the system of elected representation. This is intimately linked with the gendered division of labour and structures of decision-making. Although women predominate in welfare regimes, both as consumers and providers, they have rarely been in a position to shape their overall structures. Decision-makers, both within the state apparatus and as elected representatives, have overwhelmingly been male. In all European states except Scandinavia, women are a small minority of the legislature and in government. There has been little change in this situation since Randall's comparative study of women in politics in the mid-1980s (Randall 1987). In only one EU state (Denmark) are one third of the members of parliament women and in eight out of the twelve, fewer than one sixth of members are women. The situation is somewhat better for members of the European Parliament. In one state (Luxemburg) they are 50 per cent of the membership; over a third in two other states; and over a quarter in five more (Parliamentary Labour Party 1994).

The third element, the *gendered structure of cathexis*, refers to the pattern of emotional attachments. This is again related to the gendered division of labour in state services, with caring (women's work) still closely associated with the labour of love. It also refers to the gendered imagery in which national identities are constructed: for example, as a 'caring' state, or more explicitly as 'motherland' or 'fatherland'. Ireland, the only European Union state to be colonized, has been portrayed in nationalist mythology as female (Sales 1993); France, a colonizing power, may be seen as Marianne, while German expansionist nationalism uses imagery of the Fatherland.

The role of the family, and particularly motherhood, has been a major element in defining national identity. The Irish constitution refers to the special role of the family within the nation as well as expressly outlawing abortion. In Germany until 1977 women were explicitly defined as homemakers in the civil code (Hantrais 1994). A common metaphor is the nation as family (Kofman 1993).

Anthias and Yuval Davis (1992) argue that the way in which sexual control of women is exercised through rules governing marriage, sexuality and the family, is crucial to representing the boundaries of communities and nations. As ethnicity has been increasingly conflated with religious identity, particularly in migrant communities, these boundaries have become stronger, reinforcing male control over women. The construction of migrant cultures is a two-way process, formed in relation to a hostile and racist 'host' society. Racist discourse has used gendered imagery to evoke fears of foreigners; for example, in relation to supposed rampant sexuality.

Debate around the themes raised by Connell's analysis has perhaps been most lively among Scandinavian feminists. Evaluating women's positioning in relation to the state as citizens, clients and consumers, they have come to widely different conclusions about the nature of the gender regime. The ideology and practice of the Scandinavian social democratic system has been the most gender-equal in Europe. Hirdman nevertheless argues that the transition from a housewife to an equality contract has preserved the gender hierarchy and created new forms of segregation (cited in Duncan 1994; Siim 1993). She has been extremely critical of the invasion of the state into the private sphere of the family. Even though women are heavily involved in the public sphere, a disproportionate number of women occupy the lower echelons while power has shifted to the corporate and private sectors dominated by men. Leira (1993) speaks of women as the junior partners in the Scandinavian system, arguing that they are not integrated into the welfare state unless they behave like men. The gendered division of labour allocates the time-consuming unpaid care to women, while paid work is given more importance, and generally leads to better entitlements.

Not all feminists have come to such negative conclusions. Hernes (1987) has spoken of a woman-friendly state which 'would *allow* women to have a natural relation to their children, their work and public life'. Siim (1993), though conceding the continuing asymmetry of power between men and women, nevertheless feels that Hirdman's analysis downplays the agency of women as political actors and the mutual dependence of the family and state.

The gender contract in Hirdman's formulation is always founded on unequal initial conditions, on a basic asymmetry of power in which male domains are structurally assigned a higher value. It may well be that welfare systems and the relationship between family, community and work reflect values which were forged in the past and are still held, at least by the dominant groups. This can explain the attachment in the German welfare system to the principle of subsidiarity, or that care should be provided by the family. The dominant ideology of the bourgeois family (Pfau-Effinger 1993) was strengthened in the postwar period as the family became crucial in regaining normality after the Nazi period (Ostner 1993). Even feminists have shared the idea that women's lives, unlike men's, should be shaped by unpaid care (Ostner 1993: 95).

The state both supports and reflects gender relations in the wider society, while the ideology and practices of the state may be challenged by other groups. The Catholic Church in Ireland, for example, is a crucial support to the state in regulating sexual relations, while in France, with its explicitly secular constitution, the Church has lost its power to impose its views through legislation. The state's role in relation to gender relations may embody complex and contradictory elements. This is seen most clearly in the dualistic Anglo-Saxon model which simultaneously projects equality of opportunity and traditional family values. In spite of its projection of family values, the British state has not been able to stem the increase in single parenthood and divorce, or even the legal recognition of gay relationships in relation to parenting.

The gender regime is, however, not fixed (Connell 1990). The redefinition of women's role in social reproduction and private life in Italy in the 1970s illustrates the power of social movements to transform the prevailing ideology and practices of welfare (Bimbi 1993: 155).

TOWARDS A EUROPEAN WELFARE REGIME

We have seen that the welfare states which developed in postwar Europe included a variety of distinct regimes, with different implications for gender relations. They shared, however, a broad consensus around certain issues: that social provision should be a major and rising part of social expenditure, with a key role for the state in the provision of both public services such as health and education and social insurance. They also shared at their foundation an adherence to a male breadwinner model, which has shifted to varying degrees in the subsequent decades.

These regimes were essentially national in character, underpinned by national economic policies in which demand management to maintain employment was prominent. The crisis of profitability which developed in the 1970s, together with global restructuring of production, brought a reversal of what had seemed an inevitable rise in social provision, and a challenge to the social consensus on which European welfare regimes had been founded. As international economic integration has proceeded, individual states are less able to maintain independent policies. Jessop (1993: 22) argues 'we will witness the continuing consolidation of the "hollowed out Schumpeterian workfare state"'.

In addition, European integration is tending to bring convergence, both indirectly through the forces of the single market, and directly through European-wide policy. Cochrane (1993) argues that three features are developing across Europe. First, welfare states are becoming more mixed, with the state's role moving from provider to regulator of services. Second, social policy is becoming more linked to economic policy, with some states explicitly representing themselves as having low welfare costs in order to attract investment. Third, social policy is increasingly the responsibility of supra-national bodies, such as the European Commission.

Convergence along these lines may be detrimental to women. The first two trends point to a deepening of the neo-liberal agenda: this suggests that welfare states are likely to depend more on informal care, with increased burdens on women. An increased role for European-wide social policy could have similar implications. The model developed so far has been based predominantly on labour market regulation; it therefore excludes those outside the formal labour market, such as many women, the elderly and people with disabilities.

The relation between social benefits and labour market participation would be strengthened by moves towards the transferability of benefits for workers within the EU brought about by the Single Market. The dominance of the German economy is likely to push social insurance schemes towards the German corporatist welfare model, in which access to benefits is based on labour market status. Earnings-related schemes tend to reinforce existing inequalities, disadvantaging the low-paid, those outside the labour market and those working in the informal sector, in all of which women and migrant and ethnic minority citizens are over-represented. These developments would reinforce the trends towards increased inequalities between women, as well as between women and men.

As financial benefits move closer, there has been no harmonization on service provision, thus increasing the move in the direction of welfare system based on financial transfers rather than service provision.

While these tendencies suggest a worsening of women's position, their operation has been uneven, and it is premature to talk of convergence towards family type of female activity (Hantrais 1994: 143). Chamberlayne (1992b) argues that individual states have moved along different paths as their welfare regimes have developed, with no linear move towards convergence around a particular type of regime. Although the trends described by Cochrane have been common throughout Europe, there have been distinct policy responses which have reflected their different gendered regimes. In Britain, for example, neo-liberalism has been combined with explicitly familialist ideology and policies, whereas in France privatization and cuts in welfare have been more gender-neutral. France, too, provides evidence that, even within a corporatist model, the gender contract need not be based on a male breadwinner model. The crucial element here is the relatively high provision of childcare services.

The processes developing in Europe are complex, and the pessimistic conclusions suggested by some analyses are not inevitable. European social policy, in spite of its limitations, contains a commitment to formal equality which provides the basis for some positive policies and a framework in which progressive measures can be fought for. The entry of more Scandinavian countries into the European Union may serve to bolster the pressures for more active social policy. This may also be strengthened by the recent election result in Germany, where the Social Democrats and parties on the left increased their share of the vote. The changing composition of the European Union will be felt in the European Parliament, which is already fighting for a more active role in policy determination and for more democratic accountability. The current agenda for European integration is producing powerful tendencies towards increased inequalities and marginalization; challenging that agenda will be a crucial struggle over the next few years, as Europe moves towards closer political and economic ties.

NOTES

1 See, for example, the exception to the Maastricht Treaty negotiated by the Irish government to allow it to maintain its ban on abortion.
2 Denmark voted against ratification in its first referendum and in favour by a narrow majority in its second. In France, the majority in favour was extremely narrow.
3 In most European states, anything less than thirty-six hours per week is classified as part-time work. In Britain, 'short part-time' work predominates, which gives very limited legal protection.
4 A Ministry for the Family was created in Italy in 1993.

REFERENCES

Andall, J. (1992) 'Women migrant workers in Italy', *Women's Studies International Forum* 15: 41–8.
Anthias, F. and Yuval Davis, N. (1992) *Racialized Boundaries*, London: Routledge.
Bailey, J. (ed.) (1992) *Social Europe*, Harlow: Longman.
Barrett, M. and McIntosh, M. (1982) *The Anti-Social Family*, London: Verso.
Bennington, J. and Taylor, M. (1993) 'Changes and challenges facing the UK welfare state in the Europe of the 1990s', *Policy and Politics* 21, 2: 121–34.
Bhabba, J. and Shutter, S. (1994) *Women's Movement: Women under Immigration, Nationality and Refugee Law*, Stoke-on-Trent: Trentham Books.
Bimbi, F. (1993) 'Gender, "gift relationship" and welfare state cultures in Italy', in J. Lewis (ed.) *Women and Social Policies in Europe: Work, Family and the State*, Aldershot: Edward Elgar.
Brannen, J. and Wilson, G. (eds) (1987) *Give and Take in Families: Studies in Resource Distribution*, London: Allen & Unwin.
Bunyan, T. (ed.) (1993) *Statewatching the New Europe*, London: Statewatch and Unison.
Chamberlayne, P. (1992a) 'Income maintenance and institutional forms: a comparison of France, West Germany, Italy and Britain 1945–90', *Policy and Politics* 20, 4: 299–318.
—— (1992b) 'Models of welfare and informal care', in J. Twigg (ed.) *Informal Care in Europe*, York: SPRU, University of York.
Cochrane, A. (1993) 'Looking for a European welfare state', in A. Cochrane

and J. Clarke (eds) *Comparing Welfare States: Britain in International Context*, London: Sage.

Cochrane, A. and Clarke, J. (eds) (1993) *Comparing Welfare States: Britain in International Context*, London: Sage.

Collectivo IOE (1991) *Foreign Women in Domestic Service in Madrid*, Madrid: ILO.

Connell, R. (1990) 'The state, gender and sexual politics: theory and appraisal', *Theory and Society* 19: 507–44.

Conroy Jackson, P. (1993) 'Managing the mothers: the case of Ireland', in J. Lewis (ed.) *Women and Social Policies in Europe: Work, Family and the State*, Aldershot: Edward Elgar.

Cunningham, S. (1992) 'The development of equal opportunities: theory and practice in the European Community', *Policy and Politics* 20, 3: 177–89.

Dell'Orto, F. and Taccani, P. (1992) 'Family carers and dependent elderly people in Italy', in J. Twigg (ed.) *Informal Care in Europe*, York: SPRU, University of York.

Del Re, A. (1993) 'Vers l'Europe: politiques sociales, femmes et état en Italie entre production et reproduction', in A. Gautier and J. Heinen (eds) *Le Sexe des Politiques Sociales*, Paris: Côté-Femmes.

Doogan, K. (1992) 'The Social Charter and the Europeanisation of employment and social policy', *Policy and Politics* 20, 3: 167–76.

Duncan, S. (1994) 'Theorising differences in patriarchy', *Environment and Planning A* 26: 1177–94.

Edwards, J. and McKie, L. (1994) 'The European Economic Community – a vehicle for promoting equal opportunities in Britain?', *Critical Social Policy* 39: 51–65.

Esping-Andersen, G. (1990) *The Three Worlds of Welfare Capitalism*, Cambridge: Polity.

Fagnani, J. (1994) 'A comparison of family policies for working mothers in France and West Germany', in L. Hantrais and S. Mangen (eds) *Family Policy and the Welfare of Women*, Cross-National Research Papers, third series, Loughborough: European Research Centre, University of Loughborough.

Finch, J. (1992) 'The concept of caring: feminist and other perspectives', in J. Twigg (ed.) *Informal Care in Europe*, York: SPRU, University of York.

Franzway, S., Court, D. and Connell, R. (1989) *Staking a Claim: Feminism, Bureaucracy and the State*, Cambridge: Polity.

Ginsburg, N. (1993) 'Sweden: the social-democratic case', in A. Cochrane and J. Clarke (eds) *Comparing Welfare States: Britain in International Context*, London: Sage.

Graham, H. (1991) 'The concept of caring in feminist research: the case of domestic service', *Sociology* 25, 1: 61–78.

Grahl, J. and Teague, P. (1989) 'The cost of neo-liberal Europe', *New Left Review* 174: 33–50.

Grignon, M. (1993) 'Conceptualising French family policy: the social actors', in L. Hantrais and S. Mangen (eds) *The Policy Making Process and the Social Actors*, Cross-National Research Papers, third series, Loughborough: European Research Centre, University of Loughborough.

Hantrais, L. (1993) 'Women, work and welfare in France', in J. Lewis (ed.) *Women and Social Policies in Europe: Work, Family and the State*, Aldershot: Edward Elgar.

—— (1994) 'Comparing family policy in Britain, France and Germany', *Journal of Social Policy* 23, 2: 135–60.

Hernes, H. (1987) *Welfare State and Woman Power*, Oslo: University of Oslo Press.

Inner London Education Authority (1986) *Employment Statistics for Teachers in Secondary and Primary Schools*, London: Research and Statistics Branch, ILEA.

Jamieson, A. (1991) 'Community care for older people: policies in Britain, West

Germany and Denmark', in G. Room (ed.) *Towards a European Welfare State?*, Bristol: School of Advanced Urban Studies.

Jenson, J. (1991) 'The state and gender relations in fordist France', *Cahiers d'Encrage*, hors série.

Jessop, B. (1993) 'Towards a Schumpeterian welfare state? Preliminary remarks on post-Fordist political economy', *Studies in Political Economy* 40: 7–40.

Johnston, R. J. (1993) 'The rise and decline of the corporate-welfare state: a comparative analysis in global context', in P. J. Taylor (ed.) *Political Geography of the Twentieth Century*, Lymington: Belhaven Press.

Klug, F. (1989) ' "Oh to be in England": the British case study', in N. Yuval Davis and F. Anthias (eds) *Woman – Nation – State*, London: Macmillan.

Kofman, E. (1993) *France: Nation and Regions*, Southampton: University of Southampton Press.

Kofman, E. and Sales, R. (1992) 'Towards Fortress Europe?', *Women's Studies International Forum* 15: 23–39.

Langan, M. and Ostner, I. (1991) 'Gender and welfare', in G. Room (ed.) *Towards a European Welfare State?*, Bristol: School of Advanced Urban Studies.

Leira, A. (1992) 'Concepts of care: loving, thinking and doing', in J. Twigg (ed.) *Informal Care in Europe*, York: SPRU, University of York.

—— (1993) 'The "woman-friendly" welfare state? The case of Norway and Sweden', in J. Lewis (ed.) *Women and Social Policies in Europe: Work, Family and the State*, Aldershot: Edward Elgar.

Lewis, J. (1992) 'Gender and the development of welfare regimes', *Journal of European Social Policy* 2, 3: 159–73.

—— (ed.) (1993) *Women and Social Policies in Europe: Work, Family and the State*, Aldershot: Edward Elgar.

Liebfried, S. (1991) 'Towards a European welfare state?   on integrating poverty regimes in the EC', in G. Room (ed.) *European Developments in Social Policy*, Bristol: Bristol University Press.

—— (1993) 'Conceptualising European social policy; the EC as social actor', in L. Hantrais and S. Mangen (eds) *The Policy Making Process and the Social Actors*, Cross-National Research Papers, third series, Loughborough: European Research Centre, University of Loughborough 3, 1: 5–14.

Lister, R. (1992) 'Citizenship engendered', *Critical Social Policy* 32: 65–71.

McDowell, L. (1991) 'Gender divisions in a post-Fordist era: new contradictions or same old story?', *Transactions of the Institute of British Geographers* 16, 4: 400–19.

McLaughlin, E. (1993) 'Ireland: Catholic corporatism', in A. Cochrane and J. Clarke (eds) *Comparing Welfare States: Britain in International Context*, London: Sage.

McLaughlin, E. and Glendinning, C. (1994) 'Paying for care in Europe: is there a feminist approach?', in L. Hantrais and S. Mangen (eds) *Family Policy and the Welfare of Women*, third series, Loughborough: Cross-National Research Papers, third series, European Research Centre, University of Loughborough.

Mahon, E. (1994) 'Ireland: a private patriarchy?', *Environmental and Planning A* 26: 1277–96.

Manning, N. (1993) 'The impact of the EC on social policy at the national level: the case of Denmark, France and the United Kingdom', in L. Hantrais and S. Mangen (eds) *The Policy Making Process and the Social Actors*, Cross-National Research Papers, third series, Loughborough: European Research Centre, University of Loughborough.

Marshall, T. (1964) 'Citizenship and Social Class', in T. H. Marshall *Class, Citizenship and Social Development*, New York: Doubleday.

Maruani, M. (1992) 'The position of women in the Labour market. Trends and developments in the twelve member states of the European Community 1983–1990', *Women in Europe Supplements* 36.

Meehan, E. (1993) *Citizenship and the European Community*, London: Sage.

Mestheneos, E. and Triantafillou, A. (1992) 'Dependent elderly people in Greece and family carers', in J. Twigg (ed.) *Informal Care in Europe*, York: SPRU, University of York.

O'Connor, J. (1993) 'Gender, class and citizenship in the comparative analysis of welfare state regimes: theoretical and methodological issues', *British Journal of Sociology* 44: 501–18.

Ostner, I. (1993) 'Slow motion: women, work and the family in Germany', in J. Lewis (ed.) *Women and Social Policies in Europe: Work, Family and the State*, Aldershot: Edward Elgar.

—— (1994) 'The women and welfare debate', in L. Hantrais and S. Mangen (eds) *Family Policy and the Welfare of Women*, Cross-National Research Papers, third series, Loughborough: European Research Centre, University of Loughborough.

Parliamentary Labour Party (1994) 'Women in parliaments in the European Union', unpublished research paper.

Pateman, C. (1988) *The Sexual Contract*, Cambridge: Polity.

—— (1989) 'The patriarchal welfare state', in C. Pateman, *The Disorder of Women*, Cambridge: Polity.

Pfau-Effinger, B. (1993) 'Modernisation, culture and part-time employment: the example of Finland and West Germany', *Work, Employment and Society* 7, 3: 383–410.

Pillinger, J. (1992) *Feminising the Market: Women's Pay and Employment in the European Community*, London: Macmillan.

Pitaud, P. (1992) 'The debate on solidarities in France: towards an articulation between formal and informal sectors', in J. Twigg (ed.) *Informal Care in Europe*, York: SPRU, University of York.

Quack, S. and Maier, F. (1994) 'From state socialism to market economy – women's employment in East Germany', *Environment and Planning A* 26: 1257–76.

Randall, V. (1987) *Women and Politics. An International Perspective*, 2nd edn, London: Macmillan.

Refugee Council (1993) *Europe: Harmonisation of Asylum Politics*, Factsheet 3.

Rossiter, A. (1991) 'Bringing the margins into the centre. A review of aspects of Irish women's emigration', in S. Hutton and P. Stewart (eds) *Ireland's Histories: Aspects of State, Society and Ideology*, London: Routledge.

Sackmann, R. and Haussermann, H. (1994) 'Do regions matter? Regional differences in female labour market participation in Germany', *Environment and Planning A* 26: 1377–96.

Sales, R. (1993) 'The limits of modernisation: religious and gender inequalities in Northern Ireland', unpublished PhD thesis, Middlesex University.

Saraceno, C. and Negri, N. (1994) 'The changing Italian welfare state', *Journal of European Social Policy* 4,1: 19–34.

Siim, B. (1993) 'The gendered Scandinavian welfare states: the interplay between women's roles as mothers, workers and citizens in Denmark', in J. Lewis (ed.) *Women and Social Policies in Europe: Work, Family and the State*, Aldershot: Edward Elgar.

Taylor, D. (1989) 'Citizenship and social power', *Critical Social Policy* 26: 19–31.

Taylor-Gooby, P. (1991) 'Welfare state regimes and welfare citizenship', *Journal of European Social Policy* 1, 2: 93–106.

Twigg, J. (ed.) (1992) *Informal Care in Europe*, York: SPRU, University of York.

Waine, B. and Jordan, L. (1986) 'Women's income in and out of employment', *Critical Social Policy* 18: 63–78.

Walby, S. (1990) *Theorizing Patriarchy*, Oxford: Blackwell.

Webber, F. (1993) 'The New Europe: immigration and asylum', in T. Bunyan (ed.) *Statewatching the New Europe*, London: Statewatch and Unison.

Williams, F. (1989) *Social Policy: a Critical Introduction: Issues of Race, Gender and Class*, Cambridge: Polity.

Wilson, M. (1993) 'The German welfare state: a conservative regime in crisis', in A. Cochrane and J. Clarke (eds) *Comparing Welfare States: Britain in International Context*, London: Sage.

Yuval Davis, N. and Anthias, F. (1989) *Woman – Nation – State*, London: Macmillan.

3

WOMEN'S WORK AND EVERYDAY LIFE IN SOUTHERN EUROPE IN THE CONTEXT OF EUROPEAN INTEGRATION

Dina Vaiou

This chapter examines the prospects of women of southern Europe in the process of selective integration of countries, regions and social groups in the European Union (EU) following the Maastricht Treaty. It focuses on two interrelated issues: the experience and content of paid work and unpaid work; and the importance of the family in everyday life and in the women's strategies to cope with and change their condition.

Southern Europe is not a homogeneous area and one cannot speak of a European South as such, even though a lot of similarities can be identified among regions and countries. Nor can one speak of regions of the South as a homogeneous group, although common experiences can be found among different classes and regions. But even though significant differences do exist, more common experiences can be found among women than is evident from the 'general outlook' of various regions of the South, particularly along lines of the two issues on which this chapter focuses.

DEFINITIONS OF WORK IN THE EUROPEAN UNION

In the European Union, as in the European Economic Community before it, the question of *work* has been at the core of debate and policy-making since its constitution. In the Single European Market in the making, European peoples, the women and men in the EU regions, exist almost exclusively as factors of production, i.e. as workers. The notion of work refers only to *paid work* (or employment), however, and in the late 1980s and early 1990s it occupied a predominant position by inversion, through the prominence that unemployment acquired.

Through its scarcity, work has become the focus of attention and the central theme of two major recent Commission documents: the *White Paper on Growth, Competitiveness and Employment* (Commission of the EC

1993a) and the *Green Paper on European Social Policy* (Commission of the EC, 1993b) – both of which followed from the Maastricht Treaty. The former admits from its first page that unemployment is the reason for its existence; it discusses its roots and effects and draws guidelines for action which will lead to a considerable increase of employment by the end of the century without impeding the rates of growth and the competitiveness of European economies. Measures are proposed mainly on the supply side of the labour market. These include flexibilization of working time and employment relations, upgrading of the role of 'social partners' and improvement of human capital through training schemes. On the demand side, new areas of employment (e.g. telework, environmental improvement, services to households), tax incentives for firms to increase the number of low-skill jobs and support to small- and medium-sized enterprises are proposed to nation-states.

The Green Paper, by contrast, recognizes the centrality of work for citizens of advanced industrialized democracies, who, according to it, find it difficult to forge 'a personal and social identity when they are part of the "non-active" population' (Commission of the EC 1993b: 15). It underlines that social policy must not go into retreat in order for economic competition to recover; it discusses at length the nature of employment and unemployment and seeks to stimulate a wide-ranging debate about their future, as well as about the future of social policy in the EU. Its proposals about employment broadly follow the same lines as the White Paper, with special emphasis on 'disadvantaged groups' (including women) and on rural versus urban areas.

These documents bring again to the forefront of political, as well as academic, debate the fact that the economic benefits of a Single European Market will be quite unequally distributed among regions and social groups within the EU, hence the discussion about different phases of integration and the reference to groups who are in a disadvantaged position in the labour market.

Southern Europe, to which this chapter refers, contains many regions lagging significantly behind the EC mean on a number of indices, such as GDP per capita, sectoral shares in employment, unemployment rates and so on. Many groups of southern European women (along with migrants, youth, minorities and combinations of the above) are already disadvantaged by the fact that they live in the less developed regions of the EU. We can expect that several of these regions will continue to lag in the future, especially if regional policies, aimed to achieve cohesion and develop local potential, continue to be counterbalanced by policies to improve industrial efficiency and competitiveness at a national and European level (Amin *et al.* 1992).

The problems of deep inequalities are tackled in the EU in the context of two parallel and rarely overlapping discourses. On the one hand, the impact of European integration on regions of the EU is examined only in terms of economic indicators and performance of firms (Commission of the EC 1991b; Groupe de Recherche Européene sur les Milieux Inno-vateurs 1991). Through institutional and academic divisions of labour, the debate never hinges upon the impact of integration on individuals and groups of Europeans in the regions under consideration or upon their role in the development process. These remain the concern of other institutions and different research teams who, in their turn, usually lose sight of the importance of space and place in analysis and policy-making.

On the other hand, the problems and prospects of women's integra-tion in the Single European Market are the concern of Action Pro-grammes and other policies for equality of opportunities for women and men. Those programmes, despite their broad goals, maintain a partial focus on *women-as-workers* in the labour market and leave out of their scope important aspects of everyday life and local specificities. Moreover, they are based on assumptions about work that are not pertinent for many groups of women of the South, even in their limited definition as workers.

The pattern of work to which most research, official documents and policy-making refer at national and European levels is that of lifelong, full-time employment for a family wage, in unionized big workplaces, in industry and/or in the services, i.e. the pattern of work of the (usually male) collective mass worker (Pahl 1988). His figure lies behind most analyses and proposals while all other forms and combinations of work (usually undertaken by women) are left out, defined as 'other', less important and bound to decline. This is particularly evident in the upgraded role and dialogue of the 'social partners', promoted in recent EU documents and policies, including both the White and the Green Papers of 1993.

The notion of the social partner presupposes formal labour relations and collective organization through which workers' representation will emerge – the social partner of employers. But formal, full-time employ-ment and the model of the male mass worker have not been fully achieved, even in industrial societies; this model has never become dominant in other parts of the world; 'other' forms of work which were expected to gradually disappear have, on the contrary, persisted and possibly increased in importance. This hierarchy of forms of work – and of the workers involved – is reproduced in the recent division of forms of work into *typical* and *atypical*, launched by the EC 'Network of Experts for the Situation of Women in the Labour Market' (Meulders and Plasman 1989/1991; for a more detailed discussion see Stratigaki and Vaiou 1994). It also underlies the collection of data about employ-ment and unemployment which informs national and EC policies.

WOMEN'S PAID WORK IN SOUTHERN EUROPE

In the EU as a whole, and in southern Europe in particular, a feature of the past decade is the growing participation of women in the labour

Table 3.1 Women's activity rates, 1990

	Spain	Ireland	Greece	Portugal	Europe
Activity rate	31.9	34.5	34.9	46.8	42.4
Employment: population	24.2	29.1	30.8	43.8	37.8
Unemployment rate	24.2	15.7	11.7	6.5	11.0
Women not in paid work*	75.3	68.3	75.4	51.4	56.7

Key * Includes (unpaid) family workers, unemployed and economically inactive women as
a percentage of women of working age (15–64).

Sources Commission of the EC 1990: Tables 1, 22, 33

force, resulting from large numbers of women both joining employment
and suffering unemployment. This trend is highlighted in all official
publications of the Commission (see, for example, Commission of the
EC, various dates, *Employment in Europe*). Women's activity rates in Spain,
Italy and Greece remain much lower than the EU average, while in
Portugal they are higher, mainly due to the expansion of public services
(André 1991). But the overwhelming majority of women of the South are
registered as non-labour-force (Table 3.1). Aggregate figures, however,
reveal as much as they conceal: it is probably more correct to talk of
strongly heterogeneous, as opposed to low, participation (Bettio and Villa
1993b).

In northern European countries, women's growing activity rates have
been accompanied by an increase in the number of part-time and
temporary jobs and precarious contracts, mainly in the services
(Meulders *et al.* 1992; Commission of the EC 1992). In southern
Europe, women's increasing participation rates are connected .with the
growth of self-employment and expansion of public services (education,
health and administration). Until recently, the public sector in southern
European countries operated on a six-hour workday. This has served as
a partial substitute for part-time employment and, along with its better
social security and pension schemes, has attracted many women
employees.

Women of the South are highly concentrated in the services (Table
3.2), yet narrowly distributed across a few areas of activity (sales persons,
cleaners and caterers, primary school teachers, secretaries and nurses).
About one in four or one in five women works in industry, again in a few
branches (textiles, clothing and footwear, tobacco, toys, jewellery and
food). Practically all .those branches face technical and administrative
problems (Conroy Jackson 1990).They are also highly concentrated in
specific regions. In agriculture, women concentrate in the more traditional and less mechanized crops (olives, citrus, tobacco) where they
account for over 50 per cent of the hours of work per year. By contrast, in the more dynamic and mechanized crops (e.g. vegetable production in greenhouses, or grains) their contribution falls to under 20 per
cent of the hours of work (Vaiou *et al.* 1991).

Table 3.2 Sectoral distribution of women's employment

	Spain	Ireland	Greece	Portugal
Agriculture	13.7	9.6	35.4	22.6
Industry	16.8	23.1	17.1	25.3
Services	69.5	67.3	47.5	52.1

Source Commission of the EC 1990

A large part of women's work in the South lies beyond what national and European statistics measure and evaluate, however, out of what would classify them as economically active: agricultural work on family farms, family helpers in small businesses, industrial home-working, informal and/or temporary work in tourism, industry or personal services, irregular work in the public sector – these are areas of activity which employ a primarily female workforce.

The exact extent of these 'atypical' (by EU definitions) forms of work is impossible to assess and relevant data have to be treated cautiously. In fact, only the 'self-employed' and the 'family workers' appear in *Labour Force Surveys*, although their numbers are admittedly underestimated. Of the forms of employment where the length of the employment contract is reduced, the *Labour Force Surveys* register 'part-time' and 'fixed-term' employment.[1] Part-time is not very common or regulated in southern Europe, but fixed-term is most important in Greece, Portugal and Spain and involves many more women than men.

'Home-working' is a very widespread form of employment where the majority of workers are women.[2] It has recently been introduced in discussions of employment and the Commission has supported a working group to prepare a relevant document (Working Group on Home-working 1993). The importance of forms of employment which are defined as 'atypical' (part-time, fixed-term, casual, seasonal, work-on-call, and home-working) is therefore usually assessed indirectly, through sectoral or local studies.

A large part of 'atypical' employment falls into the realm of 'informal activities' which are, by definition, unrecorded and without contract.[3] Many women registered as housewives (non-labour-force) engage in a multitude of occupations the year round. They may be involved in farming for part of the year, in tourism during the season, in a family shop for some hours every day, or in industrial home-working, without ever gaining the status of a working person (see, among many, Hadji-michalis and Vaiou 1990a; Bimbi 1986; Comisiones Obreras 1987).

It is beyond the scope of this chapter to go into a detailed discussion of different types of such work.[4] The following examples are only an illustration of the terms of women's work in southern Europe. In agriculture, much of the seasonal labour is without contract: 55–65 per cent in the harvest, selection and packing of fruit in northern

Greece. More than 80 per cent of the workers under these conditions are women. Family farming and small family businesses in commerce and tourism – and the very common combination of all – open a significant area of women's unrecorded and sometimes non-remunerated work.

Subcontracting part of industrial production to home-workers is a means of coping with international competition for many firms in clothing, food, leather and shoes, toys, electronics and electrical equipment, paper and metal industries. Hiring young women part of the year or seasonally, when there are peaks in business (e.g. Christmas or discount periods) is a common practice in department stores, with no contracts or social security benefits. A number of services are offered on an informal basis; these include teaching courses privately, child-minding, typing, processing insurance contracts, data entry and mailing (Barthelemy *et al.* 1988; Working Group on Homeworking 1993; Vaiou *et al.* 1991).

In this context, it is important to keep in mind four points. First, despite very significant regional differences, productive activity is diffused into a very large number of small firms in all sectors (average size fewer than ten employees per firm and in commerce fewer than two) and in the increasing numbers of self-employed. Many of the firms are family enterprises relying heavily for their survival on family labour and on non-compliance with the regulative system (see also Mingione 1993). In this context 'atypical' employment finds room to develop and perhaps become the norm. Women's jobs concentrate in the very small family firms which are the majority in southern Europe. Most of these firms are hardly able to promote innovations and profit from the enlarged market and the intense competition from a unifying Europe (see, for example, Cecchini 1989; Commission of the EC 1993a). They depend on informal practices and low wages, which may ensure their short-term survival but which lead to further casualization of employment, deterioration of working conditions and reduced bargaining power in the areas where they operate.

Second, such patterns of diffused production, and the forms of employment associated with them, are not restricted to traditional or declining activities. They are also part of the restructuring trends in many dynamic branches and regions (see, among many, Mingione 1985; Hadjimichalis and Vaiou 1990b). Third Italy ('Terza Italia') is probably the best known and most widely discussed example, although in crisis since the early 1990s. The figure of the collective mass worker, mentioned earlier, is not typical in southern Europe and especially not among women. On the contrary, forms of work which are considered 'atypical' in Commission documents characterize the activity of large numbers of people, mainly women. Broadly summarizing, women in southern Europe work in declining and crisis-ridden branches of industry; they hold the majority of low-status and low-paid jobs in the services; they work in small family businesses, quite often unpaid; they are the majority of informal workers with no security or insurance; and they are seldom unionized.

Third, these features of the productive structure, along with immigration policies and 'push factors' in the places of origin, have contributed to the attractiveness of many southern European regions for foreign migrants. To a greater or lesser extent in different regions of the South, foreign migrants, a significant proportion of whom are women, are integrated in unfavourable conditions and often in the expanding informal activities where the low cost of their labour is critical. Regions in Greece, Spain and Italy which served as sources of migrants to the industrialized parts of Europe in the 1960s and 1970s, now attract large numbers of them from the Third World and from eastern and central Europe.

This is not the place to discuss the complex features of new waves of migration to and within Europe and their diverse effects on local labour markets.[5] The changing features of women's migration patterns, their problems of integration as workers and citizens, the tensions with resident women (and migrant men): these are questions which need to be studied in detail. They may qualify the picture of a unifying Europe and strongly challenge the policies promoted towards it, even in their limited scope referring to employment.

Resident or migrant women seasonal workers in Spanish or Greek agriculture, those who clean the villas where Japanese pensioners spend their holidays in the south of Portugal, part-year workers in department stores which crop up around big cities, casual labourers in tourist enterprises in all countries and home-workers in any town and region: all those women workers do not conform to the EU definitions of work. They are defined as 'other', deviations from the norm that informs EU analysis and policy.

BEYOND PAID WORK

Women's concentration in informal and atypical jobs is by no means a matter of choice. It is due, to a great extent, to a lack of alternatives (jobs in the formal part of the labour market). It is also a function of the lack of accessible and affordable social infrastructure that would enable them to look for such alternatives; in short, of the ways in which everyday life is organized. Although marked differences exist between countries and regions, women's increasing participation in money-earning labour, both registered and unregistered, has not been matched by a corresponding development of such infrastructure, nor by adjustments in the domestic sphere.

Caring and domestic labour in all of southern Europe remains 'women's work'. It is time and energy-consuming, discourages venturing into the labour market and certainly determines the conditions under which this is possible. Women bear the burden of family responsibilities without the support of dedicated partners and in the absence of adequate social infrastructure (Centre for Research on European Women *et al.* 1989). This situation not only places on women a great deal of labour that is not even considered as 'work', but it determines, to a great extent, their availability for paid work and the conditions under which they can undertake it. Restrictions are stronger among low-income women who do not have access to private caring services and usually live in areas where public social infrastructure is inadequate and/or deteriorating.

Many of these pressures are accommodated within complicated networks of mutual assistance, involving friends, neighbours and, predominantly, family whose centrality in the social fabric is a key feature of southern European life. Even though one cannot speak of a single model of the family across southern Europe (Bettino and Villa 1993a), some functions of families are important in understanding women's integration in work and everyday life; they are also an important area of difference between northern and southern Europe.

Unlike the North, in southern European countries the family has retained its role as a productive unit in all sectors and branches of economic activity, where, as already noted, small firms predominate. Families pull income together from a variety of sources and make it available to their members when they want to start a business, study, or look for a job. Through traditions of owner occupation, housing is secured for all members, but additionally, family wealth is secured through property exploitation and building. Families provide security and assistance to their members, especially in times of unemployment, and they render services to children and to the sick, the old and the disabled.

In northern Europe the 'protective net' of the welfare state has led to a progressive emancipation *from* the family of all of its members. In the South, the welfare state has been relatively undeveloped. The family serves as an alternative 'protective net', emancipation from which, by means of the labour market, is not necessarily seen as a value in itself (Bettio and Villa 1993a; Ginatempo 1992). Strong economic ties persist between members of families, particularly between parents and children.

Support by the family in all these areas (which to a certain extent substitutes for the welfare state) is essential for social integration and perhaps economic emancipation. Its 'protective net', however, goes hand in hand with the *persistence of prescriptive behaviours*, rights and duties. For it is not 'families' but specific members within them who undertake different functions. Men engage in family activities as employers or heads of micro-firms, while women (and children or youngsters) are 'family workers', more often than not unregistered and unpaid, but whose labour is a condition of existence of family businesses – and of the productive structure of many regions. Family income is pulled together by many people's labours, but other members remain men's dependants. Services to members of the family, alternatively, are rendered as a rule by women of different ages. Men's identities do not include involvement in domestic and caring labour. But such demanding and time-consuming work ties women to their homes and restricts their access to paid work and public life. It is no surprise that they are the ones who are 'available' for home-working and outwork or for 'help' in family businesses. Further, women's 'double presence' – the combination of paid work and family – relies on other women, most notably their mothers.

As employment opportunities decrease, young people often have no other choice but to engage in low-paid, precarious jobs. Work does not open a path to independence from their families. Elderly people and women also remain tied to the family, since pensions and other benefits, even when they exist, are very low. No employment or unemployment surveys can account for these discouraged potential workers. Gender and age divisions are thus quite pronounced and contribute to reproducing power relations between women and men and between young and old members as a hierarchy of positions and tasks is consolidated in the taken-for-granted area of everyday life around the family.

PROSPECTS IN THE SINGLE EUROPEAN MARKET

Official documents and the Maastricht Treaty outline a philosophy of integration and set rules within which different social agents will be acting in the years to come. In many ways the treaty is more than a constitution in that it fixes the results to be achieved by the EU (see also Lipietz 1992), thereby leaving little to be negotiated and fought for by European citizens. For example, the economic and monetary union not only places the future of the EU in a market economy, but leads back to the market whatever is not market-based – most notably social and welfare policy whose deficits can no longer be covered by central banks. Conversely, clauses about common foreign policy and common defence determine not only relations with the rest of the world but preferential alliances (with NATO and the USA), while clauses on collaboration in the domains of justice and internal affairs establish stronger barriers for non-EU citizens and strengthen the role of the police across frontiers (see also Vaiou 1993).

In the homogenizing language of the Commission, diversity is identified as the major strength of Europe, 'not only because of the rich quality of life that this brings, but also because the complex social challenges ahead call for a variety of solutions' (Commission of the EC 1993b: 14). What is less readily admitted is the fact that European diversity is associated with (and probably derives from) profound inequalities as well as hierarchical evaluation of differences. The debate and policy on work is a prime example.

Definitions of employment and unemployment and the discussion of their nature and extent in official and unofficial documents and/or in experts' reports are based on particular experiences of work, derived from the north rather than the south of the EU, and from men's rather than women's patterns of integration into the labour market. Experiences derived from different geographical settings and development histories – when invoked – are treated as 'other', less important and/or lagging. Therefore recommendations for action and policy-making are often of limited pertinence beyond their areas of origin.

The prospective benefits of dynamic firms and branches of production in the enlarged market, formal labour market regulation, unionized workers entering into social dialogue and adequate services to secure high standards of everyday life are far removed from the reality of productive structures in many regions of southern Europe and from the multiple forms, meanings and content that work has for different groups of women in the South. These forms and content do not conform to the norm which informs EU policy formulation; at the same time, they are constructed as 'other' – deviants from the dominant model of worker/social partner which underlies programmes and policies. However, patterns of work like those in which the women of the South are involved are increasingly significant even in northern Europe in the context of the flexibilization of the labour market.

It is perhaps fair to say that a minority of women who are well educated and highly skilled and have access to information and paid services will be in a position to benefit from European integration. For the majority of women of the South, however, European citizenship generalized after Maastricht is not a passport to improved conditions of work and life. In a multitude of ways (some of which I have briefly outlined) barriers are reconstituted, leaving them to bear the cost (or enjoy the benefits?) of the Single European Market in their area of residence.

Past the landmark date – 1992 – and well within the Single European Market, much of the euphoria about European integration, especially as it is projected through the Maastricht Treaty, is greatly challenged. To the extent that any debate is kept alive (e.g. in view of European elections), scepticism is articulated around a number of important issues:

- that authoritarianism and democratic deficit accrue and form the underside of European citizenship, as ever more power of decision concentrates in the hands of appointed bodies and of the growing bureaucracy;
- that in place of a professed 'common foreign policy', a battleground for economic and political penetration of individual member states is witnessed, most painfully, but by no means exclusively, in the long and bitter war in the former Yugoslavia;
- that the criteria of the Economic and Monetary Union have made inevitable the two, three, or more phases of integration, leading to the exclusion of several member states and many more

regions and revealing a clear turn away from earlier goals of harmonization;
- that unemployment is soaring and has not only led to poverty and marginalization, but has also challenged the twin foundations of postwar development in Europe, i.e. full employment and the welfare state.

In the fluidity of developments following the signing of the treaty in Maastricht, it is taken for granted that women in the South will continue to perform the same tasks and contribute to economic and social developments. But the 'comparative advantage' of their cheap labour is fast-fading: women (and men) from central and eastern Europe and the Third World are increasingly drawn into the restructuring strategies of dynamic firms as new pools of still cheaper labour. In the process of integration, 'others' are continuously constituted and deprived of the rights that a unified Europe is, in theory, promoting.

Achievement of a Europe without frontiers, in the limited part of Europe that is the EU, is effected by introducing stronger frontiers in relation to the rest of Europe and by relaxing frontier controls but retaining some barriers within the EU. In this project of selective integration, marginalization and poverty continue to be the actual reality in many regions and among many social groups. 'Fortress Europe' already contains a lot of what it seeks to leave out and continues to create social and geographical peripheries in the process of its constitution.

NOTES

1 *Part-time*: in those countries where part-time employment has a legal definition, a worker is considered part-time if his or her contract specifies that he or she should work a number of hours below the legal, conventional, or customary norm. *Fixed-term*: employment is considered to be on the basis of a fixed-term contract when it is agreed between employer and employee that the period of employment will be determined by objective criteria, such as the passage of a certain period of time for the completion of a task (Meulders and Plasman 1989/1991).
2 *Home-working*: a relatively narrow definition is used here, whereby a home-worker is someone who performs dependent work for an employer, contractor, or agent at a place of the worker's own choosing, usually the worker's own home (Working Group on Homeworking 1993).
3 *Work without contract*: the agreement between employer and employee is verbal – which gives the employer maximum flexibility to end the employment relation at any time, to avoid paying social security and other benefits and to evade labour legislation.
4 The literature on informal activities in southern Europe is vast, as the subject has been 'fashionable' for quite a while. (See, among many, Barthelemy *et al.* 1988; Bimbi 1986; Comisiones Obreras 1987; Cocco and Santos 1984; Council of Europe 1989; Hadjimichalis and Vaiou 1990a and 1990b; Recio *et at.* 1984; Sanchis 1984; Vaiou *et al.* 1991.)
5 See, for example, Calvanese 1990; Eberhart and Heinen 1992; Emke-

72 DINA VAIOU

Poulopoulou 1986; Kassimati 1991; Macioti and Pugliese 1991; Pugliese 1993; Quack *et al.* 1993; Venturini 1988.

REFERENCES

Amin, A., Charles, D. A. and Howell, J. (1992) 'Corporate restructuring and cohesion in the new Europe', *Regional Studies* 26, 4: 319–32.

André, I. M. (1991) 'Women's employment in Portugal', *Iberian Studies* 20, 1 and 2: 28–41.

Barthelemy, P., Miguelez Lobo, F., Mingione, E., Pahl, R. and Wening, A. (1988) 'Underground economy and irregular forms of employment (*travail au noir*)', programme for research and action on the development of the labour market, DG V (10 vols), Brussels: Commission of the EC.

Benigno, F. (1989) 'Famiglia mediterranea e modelli anglosassoni', *Meridiana* 6: 76–94.

Bettio, F. and Villa, P. (1993a) 'Un percorso mediterraneo per l'integrazione delle donne nel mercato del lavoro? L'esperienza italiana', in N. Giatempo (ed.) *Donne del Sud: Il Prisma Feminile sulla Questione Meridionale*, Palermo: Gelka Editori.

Bettio, F. and Villa, P. (1993b) 'Family structures and labour markets in the developed countries: the emergence of a Mediterranean route to the integration of women into the labour market', Università di Trento e Siena (mimeo).

Bimbi, F. (1986) 'Lavoro domestico, economia informale, communita', *Inchiesta* 74 (ottobre–dicembre): 22–31.

Calvanese, F. (ed.) (1990) *Emigrazione e Politica Migratoria negli Anni Ottanta*, Salerno: Dipartimento di Sociologia.

Cecchini, P. (1989) *The Challenge of 1992: Report on the Research Programme 'The Cost of Non Europe'*, Athens: Kalofolias (in Greek).

Cocco, M. R. and Santos, E. (1984) 'A economia subterranea: contributos para a sua analise quantificação no caso portugues', *Bolletin Trimestral do Banco de Portugal* 6, 1: 67–93.

Comisiones Obreras (1987) *La Mujer en la Economía Sumergida*, Madrid: Secretaría de la Mujer.

Commission of the EC (1990) *Labour Force Survey*, Luxemburg.

—— (1991a) *The Regions in the 1990s*, 4th periodic report (COM/90/609), Brussels.

—— (1991b) *The Treaty for European Unification*, Luxemburg.

—— (1992) *Employment in Europe* (COM/92/354), Luxemburg.

—— (1993a) *White Paper on Growth, Competitiveness and Employment*, Luxemburg.

—— (1993b) *Green Paper on European Social Policy*, DG V (COM/93/551), Luxemburg.

Conroy Jackson, P. (1990) 'The impact of the completion of the internal market on women in the European community', working document prepared for DG V, Equal Opportunities Unit (V/506/90-EN), Brussels: Commission of the EC.

Council of Europe (1989) *The Protection of Persons Working at Home*, Strasbourg: Council of Europe.

Centre for Research on European Women (CREW), McLoone, J. and O'Leary, M. (1989) *Infrastructures and Women's Employment*, (V/174/90-EN), Brussels: Commission of the EC.

Eberhart, E. and Heinen, J. (1992) *Central and Eastern Europe: Women Workers in the Transitional Phase*, Brussels: Commission of the EC.

Emke-Poulopoulou, H. (1986) *Problems of Migration-Repatriation*, Athens: Greek Society of Demographic Studies (in Greek).

Ginatempo, N. (1992) 'Non solo madri: condizione e identità feminile in una città del sud', *Inchiesta* 96 (aprile–giugnio): 3–21.

Groupe de Recherche Européenne sur les Milieux Innovateurs (GREMI) (1991) 'Development prospects of the Community's lagging regions and the socio-economic consequences of the completion of the internal market: an approach in terms of local milieux and innovation networks', final report, Commission of the EC, DG XVI.

Hadjimichalis, C. and Vaiou, D. (1990a) 'Flexible labour markets and regional development in northern Greece', *International Journal of Urban and Regional Research* 14, 1: 1–24.

—— (1990b) 'Whose flexibility? The politics of informalisation in southern Europe', *Capital and Class* 42: 79–106.

Kassimati, K. (1991) *Pontian Migrants from the Former Soviet Union: Their Social and Economic Integration*, Athens: KEKMOKOP.

Lipietz, A. (1992) 'Contre Maastricht parce que pour l'Europe', *Silence* 157 (decembre): 7–9.

Macioti, M. I. and Pugliese, E. (1991) *Gli Immigrati Stranieri in Italia*, Bari: Laterza.

Meulders, D. and Plasman, R. (1989/1991) *Women in Atypical Employment*, DG V (V/1426/89), Brussels: Commission of the EC.

Meulders, D., Plasman, R., Van der Stricht, V. (1992) *La Position des Femmes sur la Marché du Travail dans la Communauté Européenne, 1983–1990*, DG V (V/938/92-Fr), Brussels: Commission of the EC.

Mingione, E. (1985) 'Social reproduction of the surplus labour force: the case of southern Italy', in N. Redclift and E. Mingione (eds) *Beyond Employment*, Oxford: Blackwell.

—— (1993) 'Labour market segmentation and informal work in southern Europe', *Proceedings of the Syros Conference on 'Geographies of Integration–Geographies of Inequality' in a Post-Maastricht Europe*, Syros, Greece.

Pahl, R. (1988) *On Work*, Oxford: Blackwell.

Pugliese, E. (1993) 'Restructuring of the labour market and the role of Third World migration in Europe', *Environment and Planning D: Society and Space* 11: 513–22.

Quack, S., Papagaroufali, E. and Thanopoulou, M. (1993) *Female Repatriates from Eastern Europe: Comparing Problems and Strategies of Occupational Integration in Germany and Greece*, DG V/A/3 (V/410/94-EN), Brussels: Commission of the EC.

Recio, E. *et al.* (1984) *El Trabajo Precario en Catalunya: la Industria Textil Lanera des Valles Occidental*, Barcelona: Comision Obrera Nacional de Catalunya.

Sanchis, E. (1984) *El Trabajo a Domicilio en el Pais Valenciano*, Madrid: Instituto de la Mujer.

Stratigaki, M. and Vaiou, D. (1994) 'Women's work and informal activities in southern Europe', *Environment and Planning A* 26: 1221–34.

Vaiou, D. (1993) 'Women of the South after (like before) Maastricht?', paper presented at the conference on 'New Tendencies in Urban and Regional Development in Europe', Durham, 29–31 March.

Vaiou, D., Georgiou, Z. and Stratigaki, M. (co-ords) (1991) *Women of the South in European Integration: Problems and Prospects*, DG V (V/694/92-EN), Brussels: Commission of the EC.

Venturini, A. (1988) 'An interpretation of Mediterranean migrations', *Labour* 2: 125–54.

Working Group on Homeworking (1993) *Homeworking in the EC*, DG V (V/7173/93-EN), Brussels: Commission of the EC.

4

THE DIVERSE WORLDS OF EUROPEAN PATRIARCHY

Simon Duncan

INTRODUCTION: STRUCTURE AND DIFFERENCE IN GENDER RELATIONS

Gender divisions in which women are subordinate to men are apparently almost universal in human history.[1] At the same time, however, women in different social circumstances, and in different societies, show an immense divergence in their experiences and in their relations to men. How subordination occurs, to what degree, in what areas of life, and with what effects, all can differ substantially. Even among the relatively small group of western European countries,[2] just looking at the present and recent past, there are substantial differences in gender inequalities and gender relations (Duncan 1994a; Perrons 1994; Walby 1994). If we were to widen our scope to eastern Europe, or other parts of the world, then contrasts would be even greater.

This situation raises an explanatory dilemma. For if there is so much variation in gender relations, then how can any one concept adequately account for these differences? The discussion of patriarchy as a social structure – where men dominate, oppress and exploit women – shows this well. Gaining popularity in the 1970s, as an explanatory concept on which to base the feminist critique in social science, the notion of patriarchy later became heavily criticized as structuralist and essentialist, allowing little room for difference and for women's own capacities for action. Some commentators saw the concept as merely rhetorical or political, just drawing attention to the importance of gender inequalities, rather than playing any serious explanatory role in showing how such inequalities are created and changed. Postmodernists took this line of criticism further, arguing that there are any number of overlapping and temporary discourses of femininity and masculinity, and that these are historically and culturally viable. It is misleading, according to this view, to look for coherent, stable and repeated causal effects – a patriarchal structure – over time and culture (see Duncan 1994b for a brief review).

Nevertheless, it is undeniable that a major part of the life-experiences of both men and women, across cultures and time, is determined by the nature of their unequal relations. There are also fairly coherent, stable and repeated gendered outcomes to these relations, which often appear to have a strong causal logic. The variations in gender inequality in different western European countries are not random, unexpected, or inexplicable, but can be related to long-term social differences between them. In addition, if we are to jettison any conceptualization of explanatory structure, then gender studies risk marginalization. Without theorization of where gender inequalities come from, of how they are produced, maintained and changed, then research on gender will easily become a descriptive add-on to pre-existing gender-blind theory. The real explanatory importance of gender will have escaped again.

I take the position that we do need to develop structural concepts of gender inequality, for both these explanatory and theoretical reasons. In some ways, however, the structuralist versus postmodernist battle-lines are beside the point. For the same structure (for example, patriarchy) can work in various ways according to circumstances, and hence different outcomes will result. This allows determination without determinism, and reconciles structure with variation. This formulation follows the realist account of causality (see Sayer 1984 for a general account and Duncan 1994b for a discussion with reference to patriarchy). This position is also helpful in taking research on a further stage. For now the issue becomes one of how structures can vary, what the various outcomes are and how they are differentially caused. Comparative research becomes prioritized in this view of causality, therefore, both in describing differences and in building up an explanatory account of how social structures differentially operate. Hence the focus of this chapter. First I describe some major features of this variation in gender inequality in western Europe, focusing in turn on paid work, political motherhood and the political family. I then evaluate some current attempts to theorize this inequality, looking in turn at gendered welfare state models, notions of differentiated patriarchy and the gender contract. These theories assume a national scale. However, a major point of this chapter is that it is not adequate, in describing and understanding comparative gender inequality, to remain at the level of the national state. It is not only that there are important subnational differences in gender inequality, it is also that those gendered social processes creating such inequality will work differently in different spatial contexts – including particular regionally or locally defined contexts. The final section takes up this issue.

VARIATIONS IN GENDER INEQUALITY IN WESTERN EUROPE

Women and paid work

Taking up paid work outside the home qualifies the traditional, or at least ideal, gender roles in capitalist patriarchy of women as homemakers

and men as breadwinners. This is particularly the case if it is married
women, especially mothers, who take up paid work. Full-time jobs are
most significant in this respect, where part-time work (especially 'short'
part-time) is usually built around domestic responsibilities. This is not to
say that taking full-time paid work removes the homemaker role for
women – all the evidence shows that across Europe such women still
carry out the bulk of domestic work and caring. It does mean, however,
that they are no longer simply homemakers – they are also workers.

Figure 4.1 shows female participation rates in the paid labour force, by
age, for most European countries in 1990. Three basic patterns emerge.
First, there are countries with a bell-shaped (or reverse U) curve of
women's participation by age (Group 1 in Figure 4.1). Women are in
paid work over most of their working lives, with increasing rates for the
youngest women as training or education ceases, and decreasing rates for
the elderly as retirement takes place. This curve is the same as that for
males in all western European countries; indeed, for countries with the
highest rates in this group – Sweden and the former East Germany up to
1989 – participation rates of 80–90 per cent over most age groups are
only a few percentage points below men's. Clearly, in this group of
countries women combine paid work with homemaking and child-rearing.

The second group in Figure 4.1 is characterized by a bi-modal (or M-
shaped) curve. Women move out of paid work between the ages of 25
and 40, to concentrate on homemaking and childcare. Even though
many women do continue in paid work during this period, most work
part-time – typically women leave the labour market altogether when
their children are under school age and take part-time work when
children have reached school age. The Netherlands, with over 60 per
cent of all employed women in part-time paid work in 1989, is perhaps
archetypical. Only 25 per cent of mothers with pre-school children had
paid work, and 85 per cent of these worked at part-time jobs. Britain is
similar, with 45 per cent of all employed women, and 70 per cent of
employed mothers with pre-school children, working part-time.

If we take part-time work into account the dip in the M curve is
therefore deeper than the gross figures for overall participation. Measuring the hours of gainful employment worked shows this well (see Hakim
1994). Although there is also considerable part-time work in some of the
Group 1 countries (notably Denmark and Sweden, both with rates over
40 per cent), a substantial difference exists in the *status* of much part-time
work between Group 1 and Group 2 countries. In the former, most part-
time work is long, with rights, status and pay rates similar to those of full-
time workers. For instance, around three-quarters of Swedish female
part-timers worked 20–35 hours per week; many were mothers with
young children exercising their right to a six-hour day (see Forsberg
1994). In Britain, by contrast, half of women part-timers worked less
than ten hours, with considerably reduced employment rights, pay rates

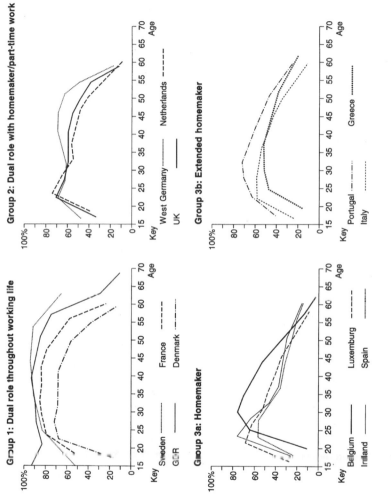

Figure 4.1 Women's labour market participation by age, Western Europe, 1990
Sources Eurostat 1992; Maier *et al.* 1992

and status compared with other workers. France, included in Group 1, is perhaps a transitional case with a residual M curve. Here, only 24 per cent of women work part-time, and most of them have 'long' part-time jobs. Indicatively, the reduction in female employment over age varies by sector in Group 2 countries (Eurostat 1992). It is in the strong economic sectors, like business services and administration, that women's withdrawal as they reach child-rearing age is most apparent, although sometimes marked M-shaped curves result as older women re-enter the sector (presumably at low grades). Female employment in the child-rearing age groups is kept up by women staying on in more traditional – and less well-paid – female sectors like clerical work and personal services. So if we were to measure women's labour market participation by income received, then the age-related M curve would be even more marked.

The third group of countries shows a single, left-hand peak reflecting a dominant homemaker role for women. Only young women show high activity rates in the paid workforce, and rates decline once and for all with marriage or childbirth. Ireland is perhaps archetypical. Part-time rates are also low, always below 20 per cent of paid women workers and as low as 10 per cent or so in Greece, Portugal and Spain. Portugal has markedly higher overall participation rates and, like Greece and Italy, slower age-related declines than others in this group. However, it is likely that this partly reflects the extended homemaker role of household employment, particularly in family farms and small businesses (see Stratigaki and Vaiou 1994). Belgium is included in this group, but may be moving towards a Group 1 curve. In all three groups, however, women's activity rates in paid work increased over the 1980s with one exception – former East Germany since 1989 (see Quack and Maier 1994 and Schmude in this book).

The question for us here, of course, is to explain these marked national groupings. First, the average level of female participation rates reflects the history of demand for labour. Those northern European countries of the postwar boom have higher average participation rates. Former East Germany – with the most severe labour shortage – had the highest of all. The peripheral capitalist economies of southern Europe and Ireland have the lowest levels.

The expansion of part-time work in Europe over the 1970s and 1980s is also pre-eminently an expansion in the demand for women's labour. This partly reflects the decline of manufacturing employment, and the rise of services – again most marked in northern Europe. Part-time work is also a strategy for employers, either to tap a latent reserve of women's labour in anticipating labour shortages, or to create a more flexible workforce. Bruegel and Hegewisch (1994) find elements of both processes in European countries. There was some national differentiation in this strategic development of part-time jobs, however. In countries with low levels of female participation, firms tended to use part-time jobs as a means of expanding labour supply, while in Britain and the Netherlands, especially, firms tended to follow a conscious flexibilization strategy. In Denmark and Sweden, by contrast, part-time work both employed relatively large numbers of men (up to 20 per cent of the total) and was decreasing. The relative strength of national labour movements, both in the workplace and in the state, may be a limiting factor on flexibilization strategies.

It is, of course, well known that women suffer from both horizontal segregation (that is, by sector and occupation) and vertical segregation (that is, hierarchically) in workplaces. A disproportionate number of women tend to end up in lower status, less secure and less paid jobs as a result (see McDowell and Court 1994). While this process is Europe-wide, differences occur in occupational sex segregation between countries, reflecting both industrial histories and the specific development of women's versus men's jobs. Furthermore, a key point for the operation of segregating processes is when women marry and have children. The salience of this point may also differ between countries and hence shape the age-related activity rates in Figure 4.1.

Feminization rates (that is, the proportion of employees who are women) by broad sector are in fact similar in all countries; women workers are concentrated in health and education in the public sector, and in personal services, clerical work and textiles/clothing in the private sector (Eurostat 1992). Not surprisingly, feminization is generally highest in Group 1 countries, and lowest in Group 3 countries, although there are interesting differences of detail. In particular, the Group 3 countries show high proportions of female workers in agriculture, while the former East Germany had an exceptionally high proportion of women workers in manufacturing. Indices of horizontal segregation reflect these trends, although in a complex manner. Some countries with low feminization show high levels of segregation between men and women at work, such as Ireland. Others, however, in Group 3 (like Greece) show low levels of segregation; this is probably because of the statistical effect of women's employment in agriculture. Conversely, countries with high feminization, noticeably Sweden, also show high indices of horizontal segregation. Here the expansion of women's employment in the 1960s and 1970s coincided with the expansion of the welfare state — new jobs, often in traditional female areas like health, were available and did not directly threaten men. Group 2 countries also show higher segregation rates (see also Perrons 1994). There are, unfortunately, no comparable data on national rates of vertical segregation, although what is available suggests that this is least severe in the Group 1 countries.

The varying national levels of women's paid employment are partly a function of the changing demand for labour, therefore. The distribution of this labour, both between sectors and hierarchically, is heavily influenced by occupational sex segregation in labour markets themselves. Given that it is no longer formally possible to pay different wages for the same job to men and women in western European countries, then it is occupational sex segregation that explains the wage gap between men and women, as shown in Table 4.1. These data indicate inequality in wage rates per unit time worked, rather than income received (the latter shows even greater inequality, as women typically work less time). Generally, the wage gap is lowest, between 10 per cent and 20 per cent, in Group 1 countries. There is little distinction between Group 2 and Group 3, however, with wage gaps of around 25–32 per cent (excepting Italy), reflecting the essentially domestic focus of women's paid work in the Group 2 countries. Participation levels may be higher than in Group 3 countries, but this is in large part a function of part-time work organized around domestic work, or re-entry (at low grades) after child-rearing age. Indeed, as Hakim (1994) shows, in terms of full-time work or hours worked there was, in fact, no increase at all in women's employment in Britain from 1971 up to the 1980s. The particularly high wage gap in Britain also reflects the importance of 'short', badly paid part-time work.

Neither the changing demand for labour by employers, however, nor occupational sex segregation in the labour market can explain why it is that women's domestic role has a different effect on labour market

Table 4.1 The wage gap: western Europe around 1990

	Manual (hourly)	Non-manual (monthly)
Group 1		
Sweden[1]	90	75/83/89
Denmark[2]	87	
Norway[3]	86	
France	81	69
Finland	78	
Group 2		
Netherlands	75	66
West Germany	73	67
UK	68	56
Group 3		
Italy[4]	83	69
Belgium	76	65
Greece	76	70
Portugal	72	71
Spain	72	62
Ireland	68	

[1] Sweden: 90% manufacturing, for non-manual 75% manufacturing, 83% local authorities, 89% central government
[2] Denmark: manual excludes iron and steel, other metals, timber
[3] Norway: manufacturing
[4] Italy: 1985
Sources Walby 1994; Eurostat 1992; Lathund om Jämställdhet 1990

participation in different countries. Figure 4.2, showing women's labour market participation by the age of their youngest child, presents this problem well. Danish and Swedish mothers with very young children (aged 0–2) show participation rates similar to those of childless mothers. (In Sweden these mothers actually have higher rates than childless mothers – abstracting from domestic responsibilities this is economically more rational; mothers need the money more and childless women can afford to train or study.) In Britain, only mothers with older children (7–15) approach the rates for childless women (although remember most of this will be part-time work, not shown in Figure 4.2). In Ireland all mothers have very low participation levels (cf. Maruani 1992). Indeed in many countries it has seemed more rational to extend male work times, or even import immigrant labour, rather than use native female labour and hence challenge in-built breadwinner/homemaker roles. Former West Germany is a good example (see Pfau-Effinger 1994). Clearly, economic rationality is gendered!

Why is it that the options and constraints facing women in combining paid work and home-work vary between these three groups of countries, as expressed by Figures 4.1 and 4.2? The most obvious explanatory candidate is the level of childcare provision. In western societies, women are left responsible for child-rearing while men take full-time paid work outside the home. If women also want paid work, where suitable jobs are available, and where men do not block this choice, then childcare provision – or its lack – will allow or deny this option. The next section describes how this varies between western European countries.

Political motherhood and public childcare provision

Motherhood plays a central part in the lives of European women and fundamentally shapes their identity and experiences. Indeed, motherhood is often seen as the true meaning of womanhood, and women without children sometimes question their very identity. Fatherhood is far less potent, both in taking up time and labour, and in shaping masculine identity. In all capitalist patriarchal societies, women normally carry out the primary parenting functions for children, reaching far beyond the biological processes of motherhood – pregnancy, delivery, breastfeeding. None the less, the mother-child relationship and forms of childcare vary significantly between European countries. Borchorst (1990) calls this political motherhood – how legislation on marriage, benefits, children's rights, maternity/paternity leave, collective childcare and so on shapes how and under what conditions motherhood is carried out, and hence also how it might be combined with other activities like paid employment. Like Borchorst, I take public childcare as an index of the varying form of political motherhood. This is because the relative availability of childcare – given that fathers do not take up a child-

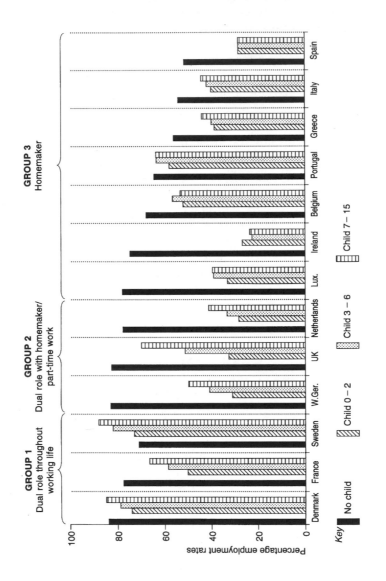

Figure 4.2 Employment of women by age group of youngest child, western Europe 1988
Source Joshi and Davies (1994)

rearing role – will directly influence the work roles of mothers with younger children.

Figure 4.3 shows the availability of public childcare in western European countries in 1989, arranged in the same groups as identified in Figure 4.1. Only in Group 1 countries is there any significant provision of public childcare to the youngest children aged 3 years and below (although Belgium, in Group 3, is a partial exception). Similarly, only Group 1 countries show much development of outside school care. This is a vitally important, although often unrecognized, component of childcare. Unless it is available for school holidays, and where schooldays are shorter than the normal working day, mothers – at least those with younger children – will find it difficult to take up full-time or long part-time work. There are also high provision rates for older, pre-school children in group 1 countries. It is in these countries, of course, that age-related rates of women's labour force participation most resemble those for men, and where the wage gap is lowest. In these cases collective childcare and paid work are part of motherhood.

Countries in Groups 2 and 3 differ much less. All show a significant development of public childcare for older pre-school children of 4 years and over. While this rate does vary considerably (with highs of 95 per cent and 85 per cent in Belgium and Italy, but only around 35 per cent in Britain and Portugal) there appears to be little correlation with female labour force participation.

This conundrum brings us to the question of the nature and extent of childcare provided, which the group estimates used in Figure 4.3 disguise. Much of the provision for older pre-school children is in fact provided in normal schools; the figures represent nursery classes or early entrance. Often they will be part-time (for instance, approximately half of Britain's provision is morning or afternoon only). The nature of childcare also varies considerably. In West Germany, most is part-time and lunch is not provided; in Sweden, childcare can cover the whole working day (sometimes night childcare is available) and includes breakfast and lunch. Similar differences affect the school day itself, varying from eight hours in France and Spain, to just four or five in West Germany – where, again, lunch is not usually provided. Whereas in Group 1 countries schooling and collective childcare allow mothers to work full-time, in West Germany childcare functions more as a support for homemaker motherhood. Mothers can better devote a short amount of time, free from children, to domestic tasks and services. Childcare is not adequate, and is not intended, as a support for employed mothers.

Disaggregating childcare rates in this way, as well as the overall lack of correlation between labour force participation and childcare levels, raises a chicken-and-egg question. Perhaps it is not so much the provision of childcare that allows mothers to take up paid work, but rather the other way round. It may be that when the definition of political motherhood includes paid work, then collective childcare is provided

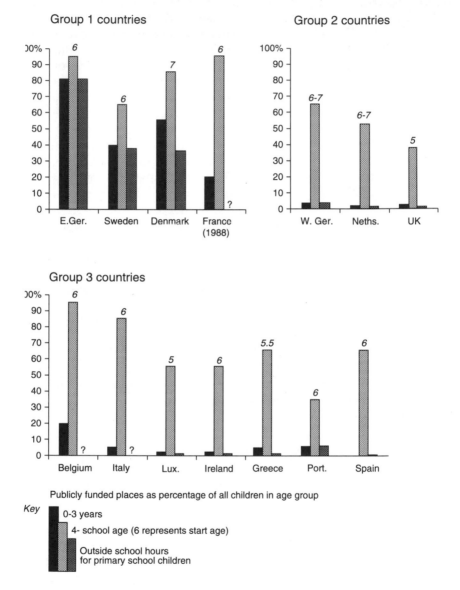

Figure 4.3 Availability of public childcare, western Europe, 1989
Sources Maier *et al.* 1991, Moss 1990

so that this can take place. Hence, in Denmark and Sweden, mothers are seen as also being workers (or perhaps workers are sometimes mothers), and childcare is provided on this basis. In other cases mothers are seen as homemakers; relatively high levels of childcare may be provided (as in Belgium or Italy) but this does not mean mothers act also as workers. In other words, the causal origin of differing female labour force participation rates, and differing versions of political motherhood, lies deep within overall societal arrangements about gender roles and identities. Each national society and each time develops a rough social consensus on what people of different sexes should do, think and be.

How is it possible to indicate differences in this gender consensus between countries? The next section attempts to gain some purchase on this problem by using indicators of actual behaviour in families, and also attitudinal data about gender and the family. Certainly the gendering of family life, as well as paid work and childcare, is central to gender inequality.

The political family – divorce and legitimacy

How men and women should relate through partnership and child-rearing, and how this relationship ascribes roles and duties, sit at the centre of gender roles. There is a core to partnership concerned with sexual exchange and intimacy, but surrounding this lies what we might call the political family . Like political motherhood, it is defined by legislation, in this case dealing with ownership rights, inheritance, taxation, benefits, dependency, sexuality and so on. Again, the shape of this political family will differ considerably between countries.

Figure 4.4 attempts to show these differences by mapping western European countries according to divorce rates and levels of birth outside marriage for 1989 (the latter index includes both single mothers and unmarried cohabitees). These are taken as indicators of the relative political importance of the traditional family in personal relationships, of whether intimacy, sexuality and child-rearing are conventionally encompassed by marriage. (Actual behaviour may differ – here we are concerned with what is seen as socially legitimate.)

A rather clear pattern of country groupings emerges. First are those countries with very low rates for both divorce and births outside marriage (Ireland, Italy, Spain, Portugal, Greece). They also have low rates of lone motherhood. In these countries relations are defined with respect to traditional marriage, where marriage is taken as a permanent relation and children are brought up within it. Indeed, in Ireland at the time of writing, divorce is illegal. These are all more peripheral capitalist countries (excepting the northern regions of Italy) with, until recently, strong religious influence on both the state and civil society. Next is a group of central European countries (the Netherlands, West Germany, Luxemburg, Belgium, Switzerland) where marriage is no longer seen as necessarily permanent, but where divorce rates are still relatively low and procreation is seen as taking place within marriage. Usually, as in West Germany, state policy is used to bolster the traditional family (Pfau-Effinger 1994). In former East Germany, in strong contrast, the state altered the nature of the political family. With high levels of female paid work and childcare provision, marriage was no longer left as the central institution for reproduction and women's economic chances. Traditional ideas of masculinity and femininity, with a strong gender coding of tasks and responsibilities, remained entrenched, however (cf. Quack and Maier 1994; Ostner 1994; and Schmude's chapter in this book). Austria, France and Norway have rates of divorce similar to the second group's, but higher rates of extra-marital births, while Finland has far higher divorce rates. Denmark and Sweden stand out with high rates for both indicators. In these countries marriage no longer provides the framework for either partnership or childbirth. Britain appears transitional.

These relative positions show some overall correspondence with what can be gleaned about attitudes to the family and sexuality. Unfortunately, a recent study is restricted to nine countries (including Northern Ireland) and does not include any Scandinavian country (Ashford and Timms 1992). However, taking a reduced version of Figure 4.4 – with France and Britain at the apex instead of Sweden and Denmark, the same broad three groupings emerge. Ireland can be seen as traditionally conservative. Here only 11 per cent of respondents in 1990 (male and female) thought marriage was outdated, only 23 per cent would approve of a conscious decision to become a lone mother, a mere 8 per cent approved of abortion on demand, just 17 per cent approved of complete sexual freedom, while 69 per cent thought that being a housewife was just as fulfilling as working for pay. Such attitudes were not, it should be noted, simply equivalent to Catholicism. Spain and Italy, for instance, recorded some of the most liberal attitudes to sexual freedom and lone motherhood. France, by contrast, seems to represent a more modernistic view of the family. Here 34 per cent of respondents thought marriage was outdated, 37 per cent approved of planned lone motherhood, 40 per cent supported abortion on demand, 30 per cent approved of complete sexual freedom while only half of respondents thought housewifing to be as fulfilling as paid work. Finally, we can distinguish conservative modernizing countries such as the Federal Republic of Germany. More liberal attitudes on some issues (34 per cent supported complete sexual freedom, 30 per cent approved of abortion on demand), were contradicted by family conservatism (only 22 per cent supported planned lone parenthood, while 62 per cent of respondents – double the nine-country average – thought that working mothers would fail to secure a relationship with their children as warm and secure as that achieved by women who stayed at home). In all countries, age of respondent was the most discriminatory factor, outweighing both sex and actual work role.

How does the centrality of the married family for intimacy and repro-

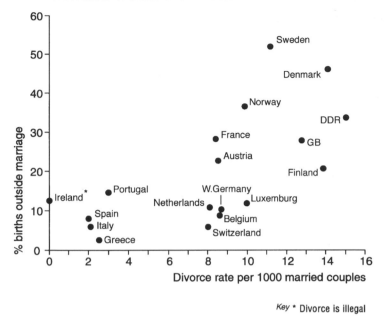

Figure 4.4 Divorce and births outside marriage, western Europe, 1989
Sources: Eurostat, national statistics

duction link into attitudes on gender divisions of labour roles more generally? Figure 4.5 shows relative country rankings on male attitudes to equal roles within the family, and egalitarian views (for both men and women) on gender roles in general, in the EC countries in 1987. It will be seen that expressed attitudes on gender roles produce a rearrangement of countries compared with Figure 4.3. Denmark remains most egalitarian, where the breakdown of marriage as a social institution is matched by egalitarian attitudes to gender roles, but it is West Germany (with Luxemburg) and Belgium that join Ireland as the most unequal. Indeed in these countries a majority of married or cohabiting men preferred that their wives did not take up paid work. Here traditional marriage in terms of intimacy and reproduction is matched by relatively conservative views on gender roles. The Netherlands has left this group, to move to a relatively egalitarian attitudinal position, while southern European countries are also more egalitarian – at least in attitudes – than their position on marriage would suggest. We may perhaps imagine that in these more modernizing countries, where older authoritarian political and religious structures are weakening or have disappeared, consensual or at least fashionable attitudes change more rapidly than relatively more embedded practice. Similarly, in West Germany, Ireland and Belgium traditional attitudes and marriage structures remain supported by state policy.

The discussion so far has focused on national variations on some

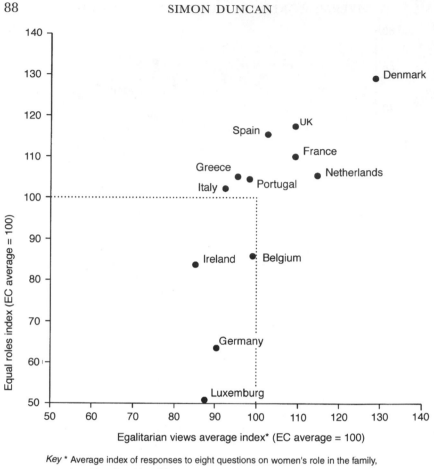

Figure 4.5 Attitudes to gender equality, western Europe, 1987
Source Women of Europe, supplement 26, 1987

important empirical indicators of gender inequality in paid work, parenting and the family. As can be seen, there are both significant
variations in these indices across western Europe, but also associations
between them, with similar groups of countries distinguished by each
measure. But how are these different systems of gender inequality
produced and what do they mean in terms of gender relations? In the
next section I discuss alternative theories of European gender systems.

THEORIZING EUROPEAN GENDER SYSTEMS

In this section I examine in turn three theoretical strands, each associated with a particular conceptual, empirical and disciplinary context.

The first strand, which I call *gendered welfare models*, are essentially feminist critiques of comparative social policy, seeking to add gender to a focus on state welfare regimes. These regimes will position women in more or less favourable ways. Ideas about strong and weak breadwinner regimes and about de-familization take this critique furthest, and in so doing provide a bridge to the second strand. I call this *differentiated patriarchy* where the issue is to develop the concept of patriarchy to take account of national differences in gender inequality. Patriarchy can vary both in its form (how male domination is maintained) and in degree (how *far* women are exploited). These ideas have been developed within the context of feminist sociology, and are particularly informed by analyses of inequality in paid work, but also of male violence. Finally, I turn to Scandinavian ideas of the *gender contract*. These have developed from a context where there have been considerable gender-neutral or even women-friendly changes in both state welfare systems and in paid work, yet gender inequality persists. Here, the concern is both with the less formal, more unseen mechanisms of gender-coding and with the establishment, and breakdown, over relatively long periods, of rough social agreements on what men and women are and what they do.

Gendered welfare models

The work of Esping-Andersen, especially his *The Three Worlds of Welfare Capitalism* (1990), represented a considerable advance in the comparative study of social policy. Most earlier work had simply described differences and similarities between various national social welfare policies. Esping-Andersen improved upon this by providing a systematic classification of three (later four) basic welfare state regimes. Crucially, this classification was not merely *ad hoc* and descriptive, but was based on analytical distinctions about what the welfare state did, how this could differ and, hence, why we would expect different outcomes in terms of social policy. Hence any particular policy, and comparisons between policies, could be related to an explanatory account of the structure of the welfare state. The explanatory analogy with the task of explaining differences in gender inequality is clear.

Esping-Andersen's starting-point was to ask how far different welfare states erode the commodity status of labour in a capitalist system (how far people are independent from selling their labour) and, as a consequence, how far welfare states intervene in the class systems. In *liberal welfare regimes* social policy is used to uphold the market and traditional work-ethic norms, with modest and means-tested benefits aimed at a residualized and stigmatized group of welfare recipients. While no one country presents a pure case of any particular regime, and countries may straddle or move between them, the USA is a type case, and Britain is rapidly moving in this direction. In *conservative welfare regimes* the preservation of status differences is central to social policy; the obsession with free markets is lacking and states intervene in a highly regulatory, although essentially conservative, way. Germany is the type case. In *social democratic welfare regimes* social policy reforms based on decommodification are extended to all classes, with equality at the highest standards rather than minimal needs. The market is de-emphasized but the high taxation necessary to finance universal welfare means that the emphasis is on avoiding problems in the first place, where every adult should be able to participate in the labour market. Sweden is archetypical. Finally, we should add the *rudimentary welfare state regimes* of southern Europe to this scheme. Here, there is no right to welfare or any history of full employment. While residualism and forced entry to the labour market remind us of the liberal model, the state can rely on surviving elements of the household subsistence economy and a large informal sector both to provide welfare and to top up employment.

Clearly, these different welfare regimes have different implications for women and men. For instance, in social democratic welfare regimes public provision of childcare will be more widely available, partly to get women into the labour market. In liberal regimes, minimal public provision will be targeted at stigmatized groups of mothers, in conservative regimes the emphasis will be on using childcare to support full-time housewives, while in the rudimentary welfare regimes little public childcare is available at all. However, Esping-Andersen's work remained gender-blind. As Lewis (1992) puts it, women disappear from the analysis as soon as they disappear from the labour market. This means that the relative importance of unpaid domestic and caring work is missed, and the various ways in which families intervene in individual positioning *vis-à-vis* markets and states are underplayed. Both of these omissions are heavily gendered, of course.

Some followers of Esping-Andersen have paid more attention to the gender-specific outcomes of welfare state regimes (for example, Ginsburg 1992). Indeed, Liebfried (1993) sees gender as central to the concept of social citizenship established in the conservative regimes (renamed Bismarckian institutional welfare states) where public policy is used to consolidate traditional male citizenship. This is opposed to the gender-neutral citizenship of Scandinavian modern welfare states. (In Anglo-Saxon residual welfare states social citizenship is being progressively reduced while on the Latin Rim of rudimentary welfare states social citizenship remains weakly developed.) Langan and Ostner (1991) go furthest in reaction to the gender blindness of Esping-Andersen's original work, by using Liebfried's variant to place the socioeconomic position of women at the centre of the classification. The *Scandinavian model* gives the appearance that women have been liberated from their dependence on men; here motherhood is partially socialized, and the tax and benefit system treats women as individual workers, so they can earn an independent income. However, most women have merely shifted from personal dependence on individual men to state dependency, while the power relations between men and women remain relatively unchanged. In the *Bismarckian model* the state is concerned to maintain the traditional model of male normal worker and female normal wife. In the *Anglo-Saxon model* women are free to choose between paid work and unpaid work at home – but as the state supports neither option women can neither enter the labour market on the same terms as men, nor can they easily maintain a strong housewife role. Finally, in the *Latin Rim model* women both cover reproduction costs in extended family systems and form the bulk of a flexible, low-wage workforce.

Adding the socioeconomic position of women in this way clearly improves the descriptive value of these welfare state models as far as gender is concerned. This is why I have called them *gendered* welfare state models. In explanatory terms, however, this solution remains extremely weak, for several linked reasons. First, gender remains essentially an optional add-on. Feminists would no doubt prefer to add it in, but it is the pre-existing theoretical core of the welfare regime model that continues to provide the explanatory dynamic. This theoretical core is firmly rooted in capital–labour divisions in a capitalist system, based around the relationship of (male, standard) workers to markets as modified by the welfare state. This is how the welfare state typologies are differentiated and where they come from. But, as Langan and Ostner (1991) point out, women are different gendered commodities from the outset, they already have different positions *vis-à-vis* markets and welfare states because of their gender. The explanatory dynamic remains gender-blind, however much gender description is added on. Second, in downplaying the original theoretical core – but in not providing an alternative – gender welfare models become descriptive rather than analytical. The naming of typologies shows this well. It is not being Anglo-Saxon, Scandinavian, etc., that leads to variations in gender inequality. Hence, and third, these categorizations have less ability to deal with variations in gender inequality than might at first appear. For example, Leira (1992) finds no overall Scandinavian model when looking at working mothers, where Norway is very different from Sweden and Denmark, while McLaughlin and Glendinning (1994) find welfare state models of little help in understanding the development of policies around paying for informal care for the elderly and disabled. As Lewis puts it, 'the categorization breaks down as soon as gender is given serious consideration' (1992: 112). Individual countries no longer fit the boxes.

In reaction to this causal weakness critics have proposed alternatives placing gender divisions in a more central theoretical position. Lewis (1992) opts for a categorization of strong, modified and weak 'bread-winner' states – Ireland, France and Sweden are type cases of each category. This approach also has the advantage of bringing unpaid work and family relations more firmly into the descriptive orbit. The weakness of this solution is that it remains a descriptive model – how and why are these states strong, modified or weak? In recognizing this limitation McLaughlin and Glendinning ask far more than simply adding to and modifying classifications of welfare regimes, instead seeking the development of 'notions of process . . . which relate to the terms and conditions under which people engage in non-market caring relations'. These would 'both intertwine with the individual-state-market commodification-decommodification processes and have their own historical trajectory'. Using a direct analogy with Esping-Andersen, they propose to label these processes 'familization' and 'de-familization', where these relate to 'the terms and conditions under which people engage in families, and the extent to which they can uphold an acceptable standard of living independently of (patriarchal) family' (1994: 65).

A combination of these two ideas, where breadwinner category describes the classificatory outcome, and (de)familization positions a formative process, is a relatively attractive option, certainly in contrast to simply gendering Esping-Andersen. None the less, this option still begs the question. What is the breadwinner state, however it may be modified? In essence, it is a colourful, but de-theoreticized, term for patriarchy. Similarly, McLaughlin and Glendinning implicitly recognize that de-familization refers to the changing place of families within patriarchal structures. This brings me to the next strand in theorizing differences in European gender systems – one which places the concept of patriarchy in a central role.

Differentiated patriarchy

Sylvia Walby's book *Theorising Patriarchy* (1990) is partly a defence of the concept of patriarchy against changes that it is essentialist, structuralist and ahistorical. She does this, however, by developing the concept so that it can embody difference. The argument incorporates two crucial steps. First, Walby draws upon realist reactions to structuralism (see Sayer 1984). Earlier grand theories of patriarchy do have problems in dealing with historical and cultural variations, but this is because they use simple base-superstructure models of causal relations. One causal element (for example, male violence or motherhood) is seen as determinant; not surprisingly there are problems in explaining variation and change in multifaceted gender inequality. In addition, the attempt to explain specific circumstances by using concepts developed at an abstract level is almost bound to be determinist and inaccurate. In other words, it is not the *substantive* notion of patriarchy, as a structured system of gender relations, which is essentialist, ahistorical and structuralist. Rather, there are faults in the way the concept has been used in constructing an explanation. Critics have confused their (correct) criticism of method with a criticism of content – the baby has been thrown out with the bathwater.

Walby's second step is to define the content, naming six patriarchal structures through which men dominate and exploit women: the patriarchal mode of production (in households), patriarchal relations in paid work, the patriarchal state, male violence, patriarchal relations in sexuality and patriarchal relations in cultural institutions. She substantiates her selection empirically by reference to previous research.

Following the realist model of explanation, these structures can now be used to develop a structural – but non-structuralist – explanatory theory which can allow both determination and variation. The six structures will be differentially developed as they interact with pre-existing situations, changes in other social structures (like capitalism) and each other. Historical and geographical outcomes will both show considerable variation and be structured by patriarchy. This reformulation of a differentiated patriarchy neatly deals with most of the criticisms raised against patriarchy as a concept of gender inequality (see Duncan 1994b for discussion). And, unlike the gender welfare models discussed earlier, this formulation has the substantial advantage that gender relations are centrally positioned in the explanatory account. Gender is not just an empirical, almost optional, add-on. It provides the central causal dynamic.

None the less, one criticism remains that Walby does not explicitly address. This is the idea that the concept of patriarchy overemphasizes the power of men and underemphasizes that of women. Women also have power; they are not merely passive victims but active social agents who change things too. However, clearly Walby's conception of an open and varying patriarchy, where how the structure works and with what results is affected by circumstances, can take account of women's power and agency more effectively than previous versions. We can tackle this issue more directly by raiding the realist account of causality once more. In realist parlance necessary relations are those that are essential to the working of a system (capital-labour relations in capitalism is a usual example; without these relations there can be no capitalism). Contingent relations are not essential to the working of the system, but may none the less be very important in how or even if it works (for instance, to carry on the example, state provision of infrastructure). Walby, in my view mistakenly, rejects this notion (see 1986: 19). For the necessary relation of patriarchy is the gender relation – there can be no patriarchy without two genders. Hence women necessarily have social position and social power, even if this is constrained and subordinate. This also means that patriarchy will necessarily vary in space and time, even neglecting all the contingencies of how it operates in different circumstances, because of these inbuilt, necessary tensions.

Walby originally developed this theory of differentiated patriarchy in the empirical context of the changing history and geography of women's paid work in Britain. Patriarchal and capitalist structures contingently interacted to produce distributions of women's paid work (Walby 1986). Towards the end of *Theorising Patriarchy* (1990) Walby applies her reformulation more generally to historical differences in the form of patriarchy, developing earlier work on the transition from private to public patriarchy. In private patriarchy, which reached a peak in the mid-nineteenth century in Britain, women are excluded from much of the public sphere and particular men, fathers or husbands, within house-holds are direct oppressors and beneficiaries. In public patriarchy, more characteristic of Britain today, women are not formally excluded from the public sphere and have achieved some freedom in their private lives. They are institutionally disadvantaged and publicly oppressed, however, while men benefit as a group – for instance, via occupational sex segregation. In passing, Walby notes the possibilities of different private/public mixes for different social groups and for different developed countries.

Walby tackles the nature of geographical differences in patriarchy more extensively in a recent article (1994). Here, she makes clear a crucial distinction between the form of patriarchy and the degree of patriarchy. The form of patriarchy refers to the relationship between the different elements of patriarchy; for instance, whether or not women have widespread access to full-time paid work over their adult lives. Public and private patriarchy represent two overall forms. The degree of patriarchy refers to the intensity or extent by which women are subordinated to men; for instance, the level of wage inequality between women and men. This scheme, according to Walby, 'creates the theoretical space to avoid ethnocentrism in comparative analysis, since a particular form is no longer necessarily associated with a particular degree of inequality' (1994: 1340). It is possible to have specific instances of public patriarchy with high degrees of patriarchy, public patriarchy with lower degrees of patriarchy, private patriarchy with high degrees of patriarchy and private patriarchy with lower degrees of patriarchy. For example, Walby discusses how Britain and Sweden both tend towards public patriarchy with high degrees of female labour market participation. However, Sweden – with a low wage gap (see Table 4.1) – has a lower degree of patriarchy. Walby also notes that the nation-state cannot be assumed to be analytically superior in describing geographical differences (as the social policy-based theories have assumed so far); ethnicity, religious affiliation, or region may be as, or more, important in understanding differences in the form and degree of patriarchy. Nevertheless it is national level data which are most easily available and Walby goes on to discuss geographical differences in the form and degree of patriarchy in western Europe in terms of nation-states.

A possible criticism of this resolution echoes a criticism of the concept of patriarchy as a whole. The various geographical combinations of the degree and form of patriarchy can be said simply to describe the varied outcomes of male domination and exploitation of women. But how have these differences arisen in the first place and, in particular, what about the actions of women in changing their circumstances? Patriarchy as a concept has been criticized as ignoring women's own social power as active agents, reducing them to the status of passive victims; certainly this geographical application of forms and degree of patriarchy does not explicitly refer to this power. I will now turn to Scandinavian theories of the 'gender contract' which more explicitly incorporate women's own actions.

Gender contracts

Scandinavian ideas of the gender contract developed in a situation where welfare states had an expressed commitment to gender equality, with explicit sex-equality programmes and where substantial women-friendly reforms and initiatives had been enacted. For instance, as indicated in my discussion of varieties in gender inequality in western Europe, because of high levels of public childcare and neutrality in the tax and benefit system, women could be both mothers and full-time workers. The overall effect of these changes, however, was to change women's lives rather than men's. Women could act in a male role, although inevitably less efficiently than men could. Activities, tasks and objects all remained resolutely gendered; gender coding of society and its equipment had not been removed – it was merely that women's space was enlarged (see Duncan 1994c; Forsberg 1994).

Ideas of the gender contract developed in this context of both substantial change in women's roles and the maintenance of gender divisions. According to Hirdmann (1988; 1990) the gender system (*genussystemet*) arranges people according to two overall rules: 1) virtually all areas of life are divided into male and female categories; and 2) this distinction is hierarchical, the male is the norm, the female is ascribed lower value. However, it is the gender contract (*genuskontrakt*) which operationalizes the gender system in specific circumstances. Each society and time develops a contract between the genders, which sets up any particular gender coding – what people of different genders should do, think and be. Note that the notion of a contract does not imply equality, men and women are not equals. Rather, the notion was developed in ironical analogy with the idea of the social democratic contract, or historical compromise, between labour and capital so beloved of political theorists where Sweden was seen as the archetypical case (Esping-Andersen is one of the latest). These political theories are gender-blind. Hirdmann notes that just as important to Swedish history was the compromise between men and women. And, like labour in capitalism, women in the gender system may be structurally subordinate; nevertheless they have substantial influence and room for action. So too, any given contract will leave numerous less defined, grey areas which become the sites of new conflict and, possibly, the origins of transition to another overall contract. The gender system will therefore show major variations in space and time, both with regard to the nature of the gender contract and to its rigidity.

Hirdmann (1990) develops this notion empirically for Sweden since the turn of the century. A *housewife contract* emerged in the 1930s and lasted into the 1960s as a response to an interwar crisis between the sexes. *Private patriarchy* had been eroded, but the position of women *vis-à-vis* marriage, domestic life and paid work was not established. The housewife contract was one attempt to resolve the competing demands for time, resources and roles made by men, married women and unmarried women (where married and unmarried women were sometimes in conflict). These conflicts, and the compromise – or gender contract – they led to, were pursued in political institutions, pre-eminently political parties, the institutional expressions of labour (such as the union movement) and capital, and the people's movements in general (e.g. the co-operative movement). This housewife contract was challenged by the circumstances of the 1950s and 1960s, leading to a *transitional phase* where women and men again voiced competing demands. A reinforced housewife contract (rather like the Bismarckian model) was not possible given the position and demands of Swedish women and eventually an *equality contract* emerged from the late 1960s onwards. While this allows women substantial gains, it too has its unsolved problems which are heightened by the circumstances of economic recession and the new political power of neo-liberalism (to become marked in the 1990s with the election of a neo-Thatcherite government). Another transitional phase may now be emerging. (See Duncan 1994b for details.)

Hirdmann does not develop these categories with reference to geographical differences at one time. However, it is clearly possible to use these categories to describe such differences; Germany shows a continuing housewife contract, Britain may be in a transitional phase from this contract and so on. Nor need this assumption be purely descriptive, for the notion of a gender contract directs attention to the question of where these contracts come from, how they are put in place and how they are maintained. Pfau-Effinger (1994) does just this for the contrasting cases of Finland and West Germany. In Finland, women are seen as independent full-time paid workers, who may also be housewives and mothers, while in former West Germany women were seen as dependent mothers and housewives who might have supplementary jobs (often part-time) at particular times. The origin of these contrasting gender contracts lies in the particular historical route from agrarian into industrial society, and the role various social groups played in this transition. Rather than being a pre-eminent actor, state policies reflect these contracts instead of initiating them.

Allocating western European countries by gender system

How might these different theorizations of European gender systems actually work out in practice? Figures 4.6 and 4.7 present alternative gender system maps of Europe based on each of the four schemata discussed here. In each case I have allocated countries to different categories following the authors involved; as they do not always designate the position of every country I have allocated the remainder, guided by the information presented in the second section of this chapter as indicated.

The allocation of countries to gendered welfare state regimes (Figure 4.6a) is simple, since it is predetermined. This process, however, also points to a major criticism of these models, as discussed above; they derive from gender-blind social explanation and do not adequately discriminate between countries in terms of gender. Ireland and Britain are quite different according to the data given above, Norway may not fit in with Denmark and Sweden, and so on. The breadwinner state model developed in reaction to these problems, and is depicted in Figure 4.6b. However, with only three categories (one with only two cases) the level of differentiation is low; in particular the strong breadwinner state category included countries as diverse as Britain, Germany and Greece. The differentiated patriarchy model, following Walby, returns us to four categories (Figure 4.7a). To some extent these are reminiscent of the gendered welfare state regime allocation, but with some greater accuracy with regard to the information presented earlier in this chapter on employment, childcare, divorce and legitimacy. Ireland now joins the Latin Rim countries and the Netherlands joins Britain in a separate category. Unlike the breadwinner model this approach allows more differentiation between the mass of strong breadwinner states. Finally, Figure 4.7b maps possible allocations by gender contract. In that Hirdmann explicitly recognizes transitional cases, this model allows the greatest level of differentiation. Of course, it would be possible to develop transitional categories in the other three schemes, although arguably Hirdmann gives the most explicit theoretical basis for doing so.

One glaring inadequacy exists in the discussion so far in both empirical differences in gender relations and their theorization. I refer to the theoretical assumption that gender relations operate at the national scale alone, and the use of *national* averages in presenting empirical differences. The crudity of Figures 4.6 and 4.7 makes this all the more obvious. For any national area well-known regional differences exist in the relative position of women. In this way mapping, as a device, is paradoxically quite illustrative. Many social policy analyses use tables with figures for each country which effectively disguise the issue of subnational variation! I will go on to discuss this issue in the next section.

GENDER INEQUALITY AND THE DIFFERENCE THAT SPACE MAKES: LOCAL AND REGIONAL

The oversimplification of spatial differences as national differences results from a conceptual neglect of the difference that space makes.

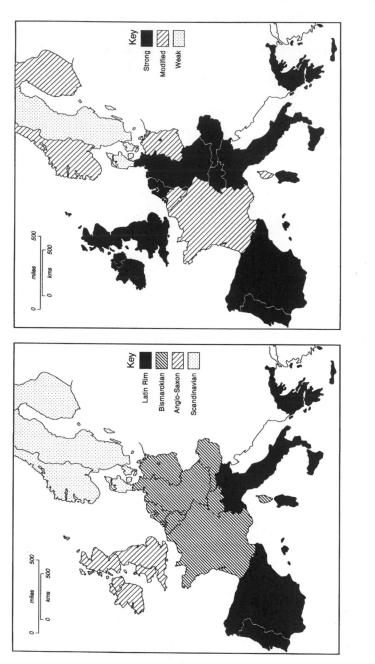

Figure 4.6 Allocating countries to gender systems: feminist social
policy models

a) Gendered welfare state regimes
Source Following Langan and Ostner 1991

b) Strong, modified and weak breadwinner states
Source Ireland, Britain, the Netherlands, E. and W. Germany, France,
Belgium, Sweden and Denmark as in Lewis 1992, Ostner 1994; other
countries allocated by author

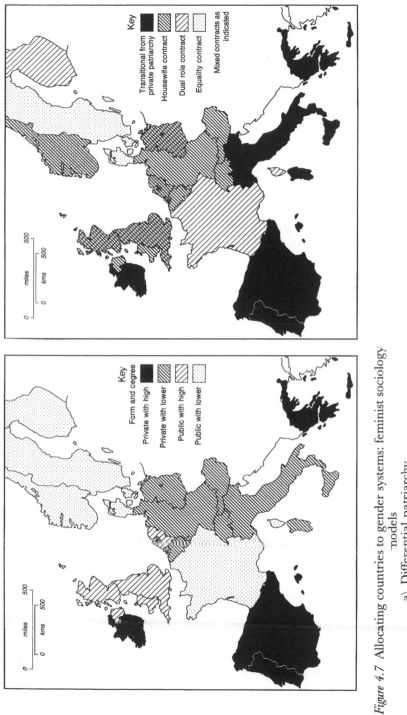

Figure 4.7 Allocating countries to gender systems: feminist sociology models

a) Differential patriarchy

Source Ireland, Britain, Denmark, Finland, Norway, Sweden as in Walby 1994; other countries allocated by author

b) National gender contracts

Source Developed by author following Hirdmann 1990

In most of social science, social processes are conceptualized as happening in a spaceless way, on the head of a pin. Theories of class, and indeed gender, are good examples. Elsewhere I have called this spatial amnesia (Duncan 1989). Unfortunately for supporters of perfect models, however, such relations are spatially differentiated. It is not just that the outcomes of social processes will be necessarily spatially differentiated as they interact with pre-existing and varying conditions (for instance, if voting patterns are heavily influenced by class position, then a pre-existing geography of class distributions will create a geography of voting). It is also that *how these processes work* will be spatially differentiated (to continue the voting example, what class means will be different in different areas, as people negotiate their identity in the contexts of local labour markets, local housing markets and social networks). Most researchers must recognize this in empirically sensitive analysis – hence the theories discussed in the previous sections – but they are left without any conceptual guide on how to incorporate spatial difference. They plump for the most immediately obvious spatial unit – the national state – and implicitly assume examples of this unit are both 1) completely separate, and 2) completely uniform. These assumptions are incorrect.

Of course, there are some good reasons for discriminating between nation-states. For this is, or at least has been, the dominant level for state intervention in societies. Welfare states, legal rules, educational and media systems have all been dominantly organized at national state level. And these spheres have obvious relevance for differences in gender relations. In the longer term, however, the nation-state can be seen as just one historical development. Currently, multistate or supra-state organizations, like the EU, are removing aspects of national specificity, and non-national institutions (such as the firms producing global satellite TV) undermine others. The history of most national states is indeed quite short, even in western Europe (e.g. Italy and Germany), not to mention the re-alignments currently taking place in much of eastern Europe. Similarly, subnational government survives (e.g. in the four constituent countries of the UK) or has been revived (notably in the German *Länder* and the *Comunidades Autonomas* of Spain). But most importantly, crucial social spheres have remained largely outside state direction – even though states may often attempt to intervene. The spatial division of capitalist labour fashions a range of variable and shifting local labour demands. These demands, because of occupational segregation, are also gendered. Similarly, definitions of masculinity and femininity will influence women's propensity to take up paid work outside the home. These definitions develop in interaction with the opportunities provided by spatial divisions of labour, leading to what seem to be enduring regional cultures of gender and work (see Sackmann and Haüssermann 1994; Duncan 1991a; and also Schmude's chapter in this book).

Focusing on national state differences alone, therefore, either will tend to restrict analysis to legal and welfare systems where national states actually do have most substantive importance, or will oversimplify through averaging, or will miss crucial social processes that operate on regional or local scales. There are two other good reasons for examining subnational state differences. First, these differences can be clearly important in their own right, in the sense of how people experience their lives and take decisions. Second, and linked to this, it is little advance to replace a theoretical structuralism at the level of systems with an empirical structuralism at the level of nation-states.

In the context of this chapter, however, there is space only to give some illustration of subnational variation in gender divisions of labour, political motherhood and the political family. Figure 4.8 presents information on dominant work roles for women within Britain in 1991 (at district council level). At a national level Britain was among the group of countries where women had a dual work role (both paid work and homemaking), but one where homemaking tended to dominate during child-rearing years (see Figure 4.1). Figure 4.8 shows the existence of marked local and regional variations. Some areas seem more completely dual-role (defined as those areas where over 40 per cent of women between 16 and 65 were in full-time work, with low levels of full-time housewives and a low part-time:full-time ratio). In other areas women's work roles were more commonly organized on a homemaker basis (over 35 per cent working as full-time housewives with high part-time:full-time ratios). Note that although urban–rural differences do emerge, there seems to be a strong regional and national differentiation. Some large urban areas are not dual-role (e.g. East London, Southampton, South Wales), whereas some small towns and rural areas in other areas (e.g. Lancashire, parts of Scotland) are. Partly, dual-role areas reflect traditions of full-time work for women (e.g. Lancashire) or extensions to this (e.g. the western crescent of economic development west of London). Nevertheless, some economically expansive areas, such as parts of this western crescent or much of East Anglia, remain homemaker areas. Similarly, traditions of full-time work have outlasted their original economic rationale. For example, the textile industry in Lancashire – where most women once worked – has virtually disappeared. Rather, the idea that women are paid workers as well as homemakers, or just homemakers, seems to be maintained by autonomous social long-standing regional definitions of gender roles and divisions of labour (see McDowell and Massey 1984; Duncan 1991a; Sackmann and Haüssermann 1994). In answering the question of how these definitions arise differentially, how they are maintained and how transitions occur, there are obvious parallels with work on differentiated national gender contracts (Pfau-Effinger 1994).

Figure 4.8 gives an indication of the varying form of patriarchy, in Britain, with reference to divisions of labour. Figure 4.9, in contrast, indicates the degree of patriarchy in paid work by measuring the wage

SIMON DUNCAN

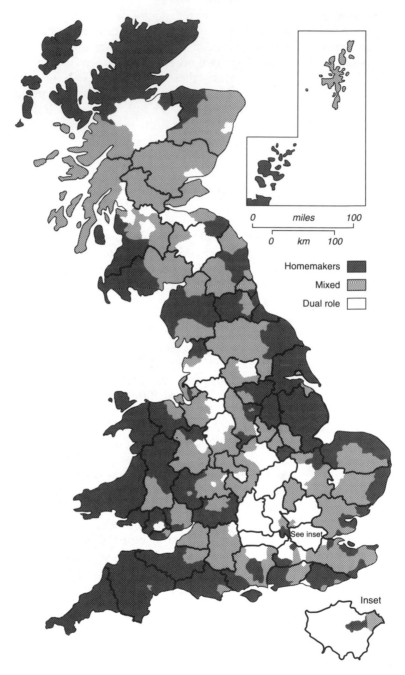

Figure 4.8 Dominant work roles for women in Britain, 1991
Source calculated from 1991 census

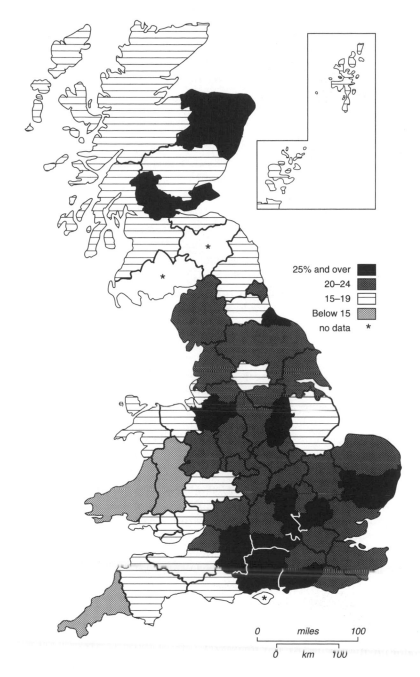

25% and over
20–24
15–19
Below 15
no data   *

0   miles   100

0   km   100

Figure 4.9 The wage gap, 1993: the percentage by which male wage exceeds
female wage
Source New Earnings Survey 1993 (average hourly, full-time wage, excl. overtime)

gap in full-time work (at county level). It will be apparent that a rather different geography of patriarchy is revealed. For in those areas where women have greater access to paid work (such as the south-east) then the wage gap is greatest. The available data do not allow inclusion of part-time work, which in Britain is particularly lowly paid. Since it is predominantly women's work, however, and is most prevalent (apart from some specialized labour markets) in the south-east, then this geography will if anything be reinforced. Those areas with the smallest wage gap, like Cornwall or West Wales, are those areas where relatively few women are in the labour market. Proportionately more employed women will be professional workers, in the education, health and social service sectors, with relatively high wages (see Duncan 1991a). Of course, this means that many more women in these areas are socially defined as housewives – as Figure 4.8 shows – although in practice many will do agricultural and tourism work. Private patriarchy is more dominant in these areas. However, where men's wages improve, women's average wages often decrease, as in the outer south-east of England with much female part-time employment. Even where women earn markedly above the female average, as in London, men earn even more above their average, so that the wage gap is retained (see Duncan 1991a; 1991b). This suggests that the entrance of women into the labour market cannot necessarily be taken as evidence of any lessening in the degree of patriarchy, even if the form has changed.

Figure 4.10 turns to the issue of political motherhood and maps the provision of pre-school day-care within England and Wales in 1991 according to local education authorities. As can be seen, the provision of day-care places in nurseries is lamentably low over the whole country (Figure 4.10a). This includes private provision – which many families are unable to afford – where public provision makes up around half of places. None the less, in a few areas at least something is available (5 per cent and over places:children) – notably some authorities in western Inner London and in Lancashire. In large areas of the country almost nothing is available. In Hampshire, no provision whatsoever was officially recorded. However, the major pre-school childcare service in Britain is provided by nursery classes and early entrance to school, which is mapped in Figure 4.10b. Here, differences are particularly marked. In some areas, provision is virtually standard for those mothers who want it, 75 per cent and over of children having places. Again, Lancashire and western London score highly, now joined by some other local education authorities such as Cleveland in the north-east. But for many rural areas and small towns in the English shires, provision remains sporadic (note also that about half of provision overall is only half-time). In other words the political definition of motherhood has a strong local state definition.[3] Almost everywhere, it is assumed that mothers of children under 3 permanently look after children, and hence

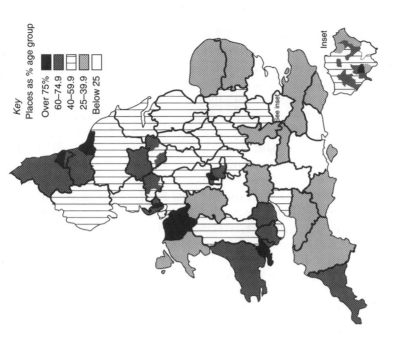

Key
Places as % age group

Over 10%
5.0–9.9
2.5–4.9
0–2.4

Inset

See inset

Figure 4.10 a) Day-care places in nurseries for under-5-year-olds, 1991

Source Department of Health, Children's Day Care Facilities; *Statistics of Education in Wales*

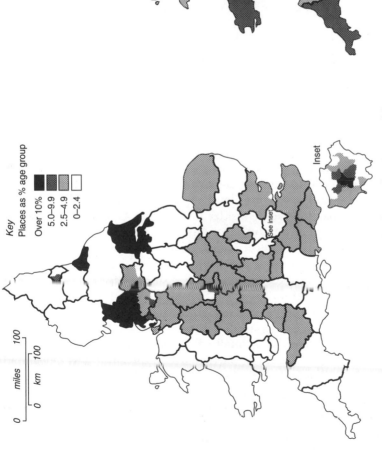

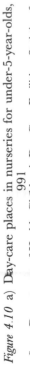

Key
Places as % age group

Over 75%
60–74.9
40–59.9
25–39.9
Below 25

Inset

See inset

Figure 4.10 b) School places for 3- and 4-year-olds, 1991

Source *Statistical Bulletin of Schools in England*; *Statistics of Education in Wales*

do not take up paid work. For mothers of older pre-school children this definition varies.

Broadly, this map of provision follows political control; traditionally Conservative areas have low levels of provision, and most high providers are run by Labour councils. However, this is only a partial correlation – many Labour areas have low provision, particularly those where a homemaker role is dominant. Similarly, some Conservative areas in west central London have a particularly high rate of provision. This is an area where non-conventional gender roles – extending to family and political life as well as paid work – are most apparent (Duncan 1991b),and there is some correlation between high day-care provision and high full-time levels of women's paid work (as in Lancashire and west London). However, as with the national level no necessary correlation exists between day-care provision levels and female labour market participation. Some areas, like Cleveland, have high provision and low participation, whereas in parts of the western crescent between London and Bristol low provision coexists with high participation.

Local state definitions of political motherhood may have limited impact on more deeply rooted regional understandings of what men and women are and what they do. Figure 4.11 turns to local and regional differentiation in the political family, comparing rates of births outside marriage at the state (*Land*) level for Germany and Austria (former East Germany is treated as one unit because of data availability). Again, substantial variation exists. Four dimensions are of importance. First, in the old West Germany note the distinction between particularly low rates in the south and west, with higher rates in the northern states. To some extent, this division will reflect religious mores – Catholic in the south, Protestant in the north. Second, there is also a division between urban modernity and rural tradition – higher rates in the city states (Bremen, Hamburg, West Berlin, Vienna) lend support to this traditional sociological view. Third, however, consider former East Germany with much higher rates than even these urban centres (so that former West Berlin appears as a relatively low-level island). Here, we may suppose that the state imposed high levels of modernity through its definition of political motherhood. There was less reason and less support for child-rearing to be specifically within marriage. Since unification, however, and the partial extension of the West German married breadwinner/homemaker model, both marriage and birth rates have collapsed (see Quack and Maier 1994). Fourth, and finally, the highest rates of all are to be found in alpine areas of Austria, particularly in the rural, less developed areas of southern Austria (Carinthia and Styria). Kytir (1992) interprets this as reflecting pre-modern patterns.[4] Rather than a birth outside marriage reflecting modernity, it reflects long-standing traditions in fairly autonomous peasant villages where decades of moral intervention by the Catholic Church and, more recently, by the national state propounding bourgeois family values have apparently had little effect. Again, this

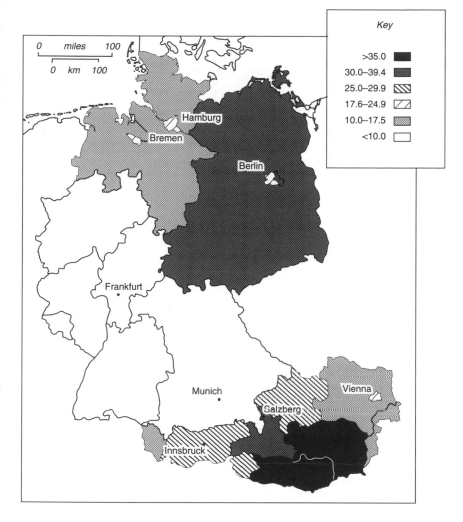

Figure 4.11 Births outside marriage, Austria and Germany 1990
Source national statistics

Austrian model appears to reflect deep-seated, regionally based household economies concerned with the regulation of population and inheritance *vis-à-vis* the control and use of resources. Hence Vorarlberg (in the
far west) reflects more the Swiss model where alternative household
economies emphasized legitimate birth (see Viazzo 1989).

Subnational variations are thus clearly important in the three crucial
aspects of gender inequality addressed in this chapter – gender divisions
of labour, political motherhood and the political family. These variations
will have substantial effects on the lives of women and men. They also
suggest that the gendered social processes causing these inequalities are

constituted at local and regional levels, as well as at national or supernational levels. The fetishism of the nation-state is inadequate.

CONCLUSIONS

In this chapter I have attempted to carry out three tasks. First, I have described the diversity of gendered systems of inequality within western Europe. There is considerable variation in divisions of labour, political motherhood and the political family, yet at the same time coherence in terms of both geographical distribution and the associations between various indicators. Second, I have discussed attempts to conceptualize European gender systems in this light, focusing in turn on gendered welfare state models, breadwinner state models (both developing from feminist critiques of social policy), models of differentiated patriarchy and the gender contract model (developing in feminist sociology). In general terms, the latter perform best as categorical systems, which would seem to reflect their theoretical superiority where gender relations form their explanatory care. Finally, I have been concerned to challenge the fetishism of the nation-state implicit in much theory and empirical work. This is a particular problem in the social policy tradition, which places national states in a privileged position. Rather, there are important subnational differences in gender inequality and, crucially, the processes producing this inequality are often constituted on subnational levels.

NOTES

1 Although note that this subordination might have been absent for some prehistorical or recent non-historical societies. See Charles 1993.
2 The boundaries of the European Union (EU) were not created in accordance with variations in gender relations, and in this chapter the seventeen countries of western Europe, including non-EU countries, are considered. In fact, the EU seems likely to include most of these non-EU states by 2000.
3 The provision of public childcare may follow social policy goals to extend educational benefits and help disadvantaged children more than any conscious labour market/gender role goal.
4 I am grateful to Rosemarie Sackmann for drawing my attention to this article.

ACKNOWLEDGEMENTS

I am extremely grateful to Jon Binnie for translating an article for me and for providing me with so many useful papers – especially his own work – to help me write this chapter. Thanks also to David Bell and the rest of the Sexuality and Space Network for their continued friendship and support; and Jan Monk and Maria Dolors García-Ramon for patiently waiting for the delivery of this manuscript.

REFERENCES

Ashford, S. and Timmons, N. (1992) *What Europe Thinks: A Study of Western European Values*, Dartmouth: Aldershot.

Borchorst, A. (1990) 'Political motherhood and childcare policies: a comparative approach to Britain and Scandinavia', in C. Ungerson (ed.) *Women and Community Care: Gender and Caring in Modern Welfare States*, London: Wheatsheaf.

Bruegel, I. and Hegewisch, A. (1994) 'Flexibilisation and part-time work in Europe', in P. Brown and R. Crompton (eds) *A New Europe: Economic Restructuring and Social Exclusion*, London: UCL Press.

Charles, N. (1993) *Gender Divisions and Social Change*, Hemel Hempstead: Harvester.

Duncan, S. S. (1989) 'Uneven development and the difference that space makes', *Geoforum* 20, 2: 131–40.

—— (1991a) 'The geography of gender divisions of labour in Britain', *Transactions of the Institute of British Geographers* 16: 420–39.

—— (1991b) 'Gender divisions of labour', in D. Green and K. Hoggart (eds) *London: A New Metropolitan Geography*, London: Unwin Hyman.

—— (1994a) (ed.) 'The Diverse Worlds of European Patriarchy', special issues of *Environment and Planning A* 26: 8 and 9.

—— (1994b) 'Theorising differences in patriarchy', *Environment and Planning A* 26: 1177–94.

—— (1994c) ' Women's and men's lives and work in Sweden: review article' , *Gender, Place and Culture* 1: 261–8.

Esping-Andersen, G. (1990) *The Three Worlds of Welfare Capitalism*, London: Polity.

Eurostat (1992) *Women in the European Community*, Luxemburg: Office for Official Publications of EC.

Forsberg, G. (1994) 'Occupational sex segregation in a women friendly society – the case of Sweden', *Environment and Planning A* 26: 1235–56.

Ginsburg, N. (1992) *Divisions of Welfare: A Critical Introduction to Comparative Social Policy*, London: Sage.

Hakim, C. (1994) 'The myth of rising female employment', *Work, Employment and Society* 7, 1: 97–120.

Hirdmann, Y. (1988) 'Genussystemet - reflexioner kring kvinnors sociala underordning', *Kvinnovetenskapligt Tidskrift* 3: 49–63.

—— (1990) *Genussystemet, in Statens Offentliga Utredningar, Demokrati och Makt i Sverige*, Stockholm: SOU.

Joshi, M. and Davies, M. (1994) 'Mothers' forgone earnings and childcare', in L. Hantrais and S. Mangen (eds) *Family Policy and the Welfare of Women*, Cross-National Research Papers, third series, Loughborough: European Research Centre, University of Loughborough.

Kytir, J. (1992) 'Zwischen vormodern und postmodern – familiengründungen in Österreich, 1984 bis 1990', *Zeitschrift für Bevölkerungswissenschaft* 18, 1: 117–33.

Langan, M. and Ostner, I. (1991) 'Gender and welfare: towards a comparative framework', in G. Room (ed.) *Towards a European Welfare State*, Bristol: School for Advanced Urban Studies, University of Bristol.

Lathund om Jämställdhet (1990) *På Tal om Kvinnor och Män*, Stockholm: Swedish Statistical Office.

Leira, A. (1992) *Welfare States and Working Mothers: The Scandinavian Experience*, Cambridge: Cambridge University Press.

Lewis, J. (1992) 'Gender and the development of welfare regimes', *Journal of European Social Policy* 2, 3: 159–73.

Liebfried, S. (1993) 'Towards a European welfare state?', in C. Jones (ed.) *New Perspectives on the Welfare State in Europe*, London: Routledge.

McDowell, L. and Court, G. (1994) 'Gender divisions of labour in the post fordist economy: the maintenance of occupational sex segregation in the financial services sector', *Environment and Planning A* 26: 1397–418.

McDowell, L. and Massey, D. (1984) 'A woman's place?', in D. Massey and J. Allen (eds) *Geography Matters*, Milton Keynes: Open University Press.

McLaughlin, E. and Glendinning, C. (1994) 'Paying for care in Europe: is there a feminist approach?', in L. Hantrais and S. Mangen (eds) *Family Policy and the Welfare of Women*, Cross-National Research Papers, third series, Loughborough: European Research Centre, University of Loughborough.

Maruani, M. (1992) 'The position of women in the labour market: trends and developments in the 12 member states of the EC, 1983-1990', *Women of Europe Supplements*, 36.

Moss, P. (1990) *Child in the European Community 1985–90*, Brussels: EC Commission.

Ostner, I. (1994) 'The women and welfare debate', in L. Hantrais and S. Mangen (eds) *Family Policy and the Welfare of Women*, Cross-National Research Papers, third series, Loughborough: European Research Centre, University of Loughborough.

Perrons, D. (1994) 'Measuring equal opportunities in European employment', *Environment and Planning A* 26: 1195–220.

Pfau-Effinger, B. (1994) 'The gender contract and part-time work by women – Finland and Germany compared', *Environment and Planning A* 26: 1355–76.

Quack, S. and Maier, F. (1994) ' From state socialism to market economy – women' s employment in East Germany', *Environment and Planning A* 26: 1257–76.

Sackmann, R. and Haüssermann, H. (1994) 'Do regions matter? Regional differences in female labour market participation in Germany', *Environment and Planning A* 26: 1377–96.

Sayer, R. A. (1984) *Method in Social Science: A Realist Approach*, London: Hutchinson.

Stratigaki, M. and Vaiou, D. (1994) 'Women's work and informal activities in southern Europe', *Environment and Planning A* 26: 1221–34.

Viazzo, P. (1989) *Upland Communities: Environment, Population and Social Structure in the Alps since the Sixteenth Century*, Cambridge: Cambridge University Press.

Walby, S. (1986) *Patriarchy at Work: Patriarchal and Capitalist Relations in Employment*, Cambridge: Polity.

—— (1990) *Theorising Patriarchy*, Oxford: Blackwell.

—— (1994) 'Methodological and theoretical issues in the comparative analysis of gender relations in western Europe', *Environment and Planning A* 26: 1339–54.

5

AN EQUAL PLACE TO WORK?
ANTI-LESBIAN
DISCRIMINATION AND SEXUAL
CITIZENSHIP IN THE
EUROPEAN UNION

Gill Valentine

INTRODUCTION

Lesbians and gay men are, Jon Binnie (1994) has argued, Europe's invisible citizens – overlooked or deliberately excluded from much of the legislation forging the European Union (EU). But despite being left off the European Union's (EU) political agenda, sexual dissidents are seizing the initiative to establish Europe as a stage for transnational lesbian and gay organizations. In 1991 a European Lesbian Conference was held in Barcelona. It was followed in July 1992 by the staging of the first ever Europride in London. Over 100,000 lesbians and gay men from across Europe marched in the largest ever exhibition of European lesbian and gay consciousness. Since then similar festivals have been held in Berlin (1993) and Amsterdam (1994) (Binnie 1994).

The time has now come, according to the leading British gay activist Peter Tatchell, for Europe's invisible citizens to formalize their transnational European networks, such as the International Lesbian and Gay Association, into a co-ordinated Federation of Lesbian and Gay organizations. This would allow European groups to join forces and resources in order to run political campaigns to advance lesbian and gay interests across member states (Andermahr 1992). He claims that 'it is through collective solidarity, overriding national boundaries and sectional interest, that we have our surest hope of eventually winning equality for the 32 million lesbian and gay citizens of the EC'[1] (Tatchell 1992: 75, quoted in Binnie 1994: 3).[2]

If Tatchell's Euroambitions are to be realized, European lesbians and gay men with very different geographies, histories and cultures need to identify specific issues around which they can mobilize to pursue common aims.[3] For lesbians, discrimination in the workplace is perhaps the most significant problem shared by Euro dykes.

In the last forty years women's participation in the European and

North American labour markets has increased dramatically and, consequently, so has the number of lesbians in employment. Numerous surveys
show that most lesbians are in paid work or are seeking employment with
only 1 per cent claiming to be full-time homemakers (Jay and Young
1977; Hall 1986). Hall (1986) argues that this focus on paid work 'may
be related to the unfeasibility, for lesbians, of other options. Unable to
marry legally, lesbians are not entitled to share their partner's assets [in
most European States]. Women earn significantly less than men and are
less able to support someone else' (Hall 1986: 60). Given this reliance on
paid employment, equality in the workplace is therefore essential for the
economic survival of gay women across Europe.

This chapter[4] begins by examining European lesbians' shared experiences of discrimination in the workplace. A key aim of European
integration is to facilitate the free movement of workers across the EU.
The chapter therefore goes on to explore how differences in the national
legislation of member states with regard to discrimination and the rights
of citizens may impinge on the free movement of lesbian workers. The
geographical specificities of oppression in the workplace is a theme that
is developed further in the conclusion. It considers different cultural
pressures that silence lesbians at work even where the law is tolerant
of lesbian life-styles.

COMMON OPPRESSIONS: ANTI-LESBIAN
DISCRIMINATION IN THE EUROPEAN
WORKPLACE

Anti-gay discrimination is prevalent in European workplaces. A survey
carried out in 1984 by the UK group Lesbian Employment Rights found
that 151 out of 171 gay women questioned in London had experienced
some form of anti-lesbianism in the workplace (Taylor 1986). Ten years
later, in 1994, Stonewall, the UK lesbian and gay equality campaign
group, carried out a similar survey of lesbians' and gay men's experiences. One sixth of the 1,873 respondents claimed to have experienced
discrimination, one fifth suspected that they had been discriminated
against; 8 per cent had been sacked because of their sexuality; half
said they had experienced harassment and a quarter said they were
too afraid to apply for certain jobs or to specific employers.

Although there are more widely reported cases of discrimination
against lesbians in the UK than in other European countries, this is
not to suggest that British lesbians actually experience more harassment[5]
(Betten 1993). Rather, as the report on *Sexual Discrimination at the Workplace*
(1984) produced by the Italian MEP Vera Squarcialupi demonstrates,
discrimination at work is an experience shared by lesbians across
Europe.

The Squarcialupi report found that 'although most European legislatures are now more tolerant with regard to homosexuality between

adults, they do not provide any protection against discrimination' (Betten 1993: 343). It drew particular attention to the way that vetting and security checks are used to weed out lesbians and gay men during recruitment for public office throughout Europe and demonstrated that lesbians are systematically excluded from some forms of employment; for example, military service. Similar findings were also reproduced in *The Iceberg Report, Anti-homosexual Discrimination in Europe 1980–90* (University of Utrecht 1991). This demonstrated that lesbians' and gay men's prospects of employment and/or promotion are frustrated by anti-gay laws in European military services. For example, Italy cites homosexuality as grounds for 'unfitness to serve'; and although the Federal Republic of Germany does not exclude sexual dissidents from the services, they are not allowed to hold senior or educational positions (Waaldijk 1993). Between 1987 and 1989, 196 men and women were discharged from the UK forces for being lesbian or gay (Hugill 1992; Simpson 1992). And in 1994 the UK Ministry of Defence admitted compiling a 'lesbian index' on its central *criminal* record computer, listing women who have been investigated by the military police as suspected lesbians and the names of their lovers and acquaintances (*Diva* 1994a). These files are available to other organizations including the police and British Telecommunications (*Pink Paper* 1994a).

Teaching is another occupation that reproduces compulsory heterosexuality (Khayatt 1992). Khayatt argues in her book *Lesbian Teachers: An Invisible Presence* that female teachers' roles have been defined in order to create an image of gentle and decent womanhood and that by acting *in loco parentis* as the nurturing and caring purveyors of morality to young people, women teachers are assumed to be heterosexual. A lesbian teacher is therefore almost a contradiction in terms (Leathwood 1994) and is commonly perceived to be likely to corrupt children in her care physically or ideologically (Squirrell 1989).

In 1980 a Belgian deputy headteacher featured in a television programme in which she discussed the way that she felt her career had been hampered by her lesbianism. She was promptly sacked by her employer, the provincial authority, because she had 'deliberately challenged the Provincial authorities by arguing that her homosexuality had been an obstacle to her being appointed as a headmistress' and that she had 'insinuated that it was just as dangerous to have entrusted the headship to two men, thus challenging the integrity and moral sense of her immediate superiors' (Waaldijk 1993: 112). The teacher's appeal against her dismissal made to the European Council of State was thrown out because it was argued that a civil servant should not make unjustified criticism of a superior. The European Commission of Human Rights also refused to entertain an application from her (Waaldijk 1993).

Discrimination against lesbians at work also extends to workplace culture and the provision of work-related benefits available to employees' 'families' (Betten 1993; Waaldijk 1993). Whilst the heterosexual 'family' is seen to complement working organizations by 'providing continuity and the rest and recreations workers need to be productive, the gay lifestyle is not perceived to be stable or to offer the same restoratives' (Hall 1989: 126). And so whilst many European organizations take a paternalistic approach to their heterosexual workers by, for example, providing health insurance and other benefits for their families, lesbians' partners and their children are usually not eligible for the same perks. European employers therefore both organize and reproduce heterosexual hegemony in the workplace (Valentine 1993).

Heterosexual desire and constructions of heterosexual attractiveness position men and women differently within workplace culture (McDowell 1995). Gutek (1989) argues that women at work are perceived to be inherently sexual in appearance, dress and behaviour. She says: 'Because it is expected, people notice female sexuality, and believe it is normal, natural, an outgrowth of being female' (Gutek 1989: 60). Correspondingly, although men are usually perceived to be asexual in appearance, they use 'sex' at work to tease, flirt with and manipulate women. A heterosexual performance is such an intrinsic part of many client–customer interactions that some employers even train their staff to use scripted heterosexual exchanges when dealing with clients (Leidner 1991). Through self-surveillance and the watchful gaze of others, women's bodies are effectively disciplined in many workplaces in order to enforce these gender and heterosexual identities in the maintenance of patriarchy (Fiske 1993). Lesbians therefore have to be particularly vigilant at work in order to manage how their appearance is read because, as Hall (1989: 127) argues:

> the sheer weight of dominant cultural attributions that lesbians must carry, if their orientation is known, renders them unavailable for the myriad and quickly shifting micro-projections necessary to maintain and elaborate the male narrative of the self.
>
> (Hall 1989: 127)

Whereas most heterosexuals take for granted their freedom to express their sexuality publicly, using their relationships and sexual experiences as common currency in workplace conversations (Pringle 1988), many lesbians, conscious of their vulnerability to discrimination from employers and workmates, conceal their sexuality or fabricate heterosexual experiences. This often involves maintaining a rigid separation of home and work; for example, by avoiding colleagues' housewarming parties and social events. Research (Valentine 1993) suggests that the creating of this artificial public–private split makes many lesbians feel 'out of place' in workplace culture and find it difficult to network with colleagues and establish authentic friendships with workmates. Missing out on work 'gossip' and being perceived as 'not a good team player' can in itself be a significant barrier to promotion.

A lesbian gynaecologist told the UK paper the *Guardian*:

I have to shroud my social life in mystery, avoid awkward questions
and often tell lies. I am not paranoid; I'm quite sure that a majority
of my colleagues would strongly disapprove and a small minority
would be extremely hostile.

(Jury 1993: 11)

Lesbians who are less circumspect about their sexuality may escape
formal discrimination by employers but still have to endure the homophobia of colleagues. In *Coming out of the Blue* (Burke 1993), a book based
on interviews with UK lesbian and gay police officers, a lesbian detective
sergeant recounts how graffiti about her were written all over the
station's toilets and how her colleagues presented her with a vibrator
at Christmas. As Evans argues: 'Whether revealed or not, employment
practices constrain and attribute lower value to the labour of sexual
minorities, reinforcing dominant legal and moral sexual standards'
(Evans 1993: 40).

These examples of anti-gay practices in the workplace across member
states confirm the need for the EU to adopt anti-gay discrimination
measures. Following the publication of Italian MEP Vera Squarcialupi's European report on *Sexual Discrimination at the Workplace*, lesbian
and gay issues did temporarily establish a toe-hold on the European
political agenda, when the European Parliament (EP) adopted a resolution which requested the Commission of the European Community (EC)
to submit proposals to ensure that:

- no cases arise in the member states of discrimination against
  homosexuals with regard to access to employment and working
  conditions
- steps be taken to induce the World Health Organisation (WHO)
  to delete homosexuality from its International Classification of
  Diseases[6]
- member states be invited to provide a list of all provisions in their
  legislation which concern lesbians and gay men
- and to identify on the basis of such lists, any discrimination with
  regard to employment, housing and other social problems by
  drawing up a report, pursuant to Article 122 EEC treaty.

(Betten 1993: 344)

But the EP's initiative came to nothing in the face of the reluctance of
the Commission to take any action. Although the then EC Commissioner welcomed the EP's recommendations, he also expressed the
opinion that 'the EEC [European Economic Community] Treaty does
not contain any provision which would authorize legal measures to
abolish discrimination [against lesbians and gay men] with regard to
access to the working place and working conditions' (Betten 1993: 344).
This, despite the fact that in 1976, faced with the same lack of legal
authority, the Commission had seen fit to establish a *Directive on Equal*

Treatment of Men and Women in the Workplace. The Commissioner did, however, express more sympathy towards sexual dissidents who were actually dismissed on the grounds of sexuality, saying that the Commission was preparing a Directive to unify grounds for dismissal across national states, and that sexual orientation would not be defined as an acceptable ground for dismissal (Betten 1993).

In 1989 the EP passed a Resolution on the *Community Charter of the Fundamental Rights and Freedoms of EC Citizens* calling for equal protection for workers regardless of, amongst other things, sexual preference; but in the same year a proposal to include anti-lesbian and gay discrimination in the Parliament's *Charter on Fundamental Rights and Freedoms of EC Citizens* was turned down (Betten 1993). Similarly, although *The Code of Practice on Measures to Combat Sexual Harassment* adopted by the EC Commission on 27 November 1991 explicitly states that harassment of lesbians and gay men falls within the definition of sexual harassment, the *Recommendation on the Protection of the Dignity of Women and Men at Work*, adopted on the same date, does not (Betten 1993).

The Commission's justification for its refusal to offer lesbians and gay men the same full protection in the workplace offered to women employees is a classic example of a chicken-and-egg argument. As Andermahr explains:

> Although EC Treaties uphold general human rights, none of them specifically mentions the rights of lesbians and gay men. This omission allows the Commission to avoid legislating on what is seen as a controversial issue. Thus the main obstacle to securing lesbian and gay rights through the EC is its alleged incompetence to address the issue of anti-homosexual discrimination. It consistently argues that it is governed by EC Treaties which omit explicit reference to lesbians and gay men. As gay activists point out, this constitutes an extremely conservative interpretation of the EC treaties, following the letter rather than the spirit of their human rights declarations and demonstrates a lack of political will on the EC's part to advance lesbian and gay rights.
>
> (Andermahr 1992: 112)

When a dispute arises about whether a group needs/is entitled to equality, lesbian academic Angela Wilson argues that Raz's principle of harm should be adopted as a practical guideline to judge whether positive action such as anti-discrimination legislation is required (Wilson 1993). 'For Raz harm includes not only physical harm, but also coercion, manipulation and even socially harmful attitudes such as racism [or homophobia]. Moreover, harm at the hands of society or government involves, according to Raz, "depriving a person of opportunities or of the ability to use them"' as a '"way of causing him [*sic*] harm . . . harm to a person may consist . . . in frustrating his pursuit of the projects and relationships he [*sic*] has set upon"' (Raz 1986: 413 in

Wilson 1993: 185). Given the evidence of the Squarcialupi report and other European surveys, it is clear that in these terms the European Parliament and the Commission do have a duty to act to establish equality for lesbians in the workplace under the law.

Phelan has argued that 'Lack of protection against social and economic harassment demonstrates, not that the liberal state is failing to live up to its standards, but that the standards leave huge loopholes in the most intimate, most defining areas of our lives' (Phelan 1989: 18, quoted in Wilson 1993: 186). Andermahr is more forthright in her criticism of the European legislatures when she claims that 'Lesbian rights are not a secondary issue; they are part of the primary issue facing all Europeans: that of enabling citizenship for all' (Andermahr 1992: 115).

Ironically, perhaps, whilst European lesbians are establishing their *commonalities* of oppression and organizing transnational political networks to campaign for legal protection, as the next section of this chapter outlines, the most promising strategy to lever equal rights legislation out of the EU appears to be to expose the fact that its goal of harmonization is being contradicted by *differences* between member states' legislation on lesbian and gay rights.

GEOGRAPHIES OF DIFFERENCE: SEXUAL CITIZENSHIP AND THE MIGRATION OF LESBIAN WORKERS IN THE EUROPEAN UNION

Even within seemingly homogenous developed societies of North West Europe, there are significant national (and local) differences in the way sexuality is regulated by the state. This is not just true of nation-states, as significant regional and local differences also exist within nation-states.

(Binnie 1994: 2)

European geographies of sexual citizenship have been shifting throughout the 1980s and 1990s. Some states such as Denmark, the Netherlands and the Irish Republic have taken a more liberal line towards lesbians and gay men; and anti-discrimination legislation is forthcoming in Belgium. Other states, notably the UK, have introduced more restrictive legislation banning the 'promotion of homosexuality', whilst lesbians and gay men are barely visible in some of the East European states that are eager to join the EU (Binnie 1994). In particular, marked differences exist between the protections afforded to lesbian workers under anti-discrimination laws by different member states.

In France it is unlawful to dismiss or not employ a person on the grounds of the 'habits' of the worker (which covers sexual orientation) whilst in the Netherlands it is a criminal offence 'in the performance of a public office, a profession or business to discriminate against persons on account of their heterosexual or homosexual orientation' (Waaldijk

1993: 106). Whilst Sweden has explicitly prohibited anti-lesbian and gay discrimination in the workplace since 1987, it was only in 1993 that the Irish Republic amended its Unfair Dismissals Act to prevent workers being fired solely on the grounds of sexual orientation. At the time of writing, the Irish Government was committed to bringing in an amended Employment Equality Act and a new Equal Status Bill to expand significantly the legal rights of Irish sexual dissidents (Rose 1994).

The picture is equally complex in relation to the rights extended to lesbian migrant workers and their partners by different European states. Whilst heterosexual married couples are basically extended the same rights throughout member states, the legal status of unmarried couples, specifically of lesbian and gay partnerships, varies widely geographically. For example, in the Netherlands unmarried couples (including same-sex couples) have equality with married citizens if they can prove they have been living in a stable relationship for three years. In Spain a law has been passed permitting the civil registration of lesbian and gay couples (Murray 1994) and in Denmark registered partnership laws confer on same-sex relationships similar rights to those enjoyed by heterosexual couples through marriage.

In most of the rest of Europe, however, unwed couples have no immigration status. Whilst unmarried heterosexual couples can go through a quick ceremony of convenience to get round this barrier to mobility, lesbians have no such option. 'Insofar as marriage between persons of the same sex is not allowed by most legislation, acquisition of citizenship by way of marriage is impossible for lesbian and gay couples of different nationalities' (Tanca 1993: 280, quoted in Binnie 1994: 6). In May 1992, Cornelia Scheel, daughter of a former West German president, and Hella Von Sinnen, a well-known German comedian, applied to be married in a bid to win recognition for same-sex relationships. Their stand was followed three months later by a nationwide 'Register Office Action', as part of a continuing campaign by German lesbians to win legal equality with heterosexuals (Lorenz 1992). In the UK the Lesbian Avengers have also demonstrated at Heathrow airport to draw attention to the predicament faced by lesbian couples of different nationalities because of discriminatory partnership legislation (*Diva* 1994b).

Thus, for example, whilst a Danish lesbian would face the prospect of separation from her partner if she migrated for work to the UK, her British counterpart would be able to seek work in Denmark without breaking up her 'family'. Inequality before the law can therefore prevent lesbians migrating from countries with a more liberal attitude to those that adopt a more restrictive line. Likewise, different legal geographies can hinder those who want to move from a restrictive environment to a more tolerant state because their curriculum vitae may be blighted by a 'criminal' record or the fact that they have been dismissed from military service on the grounds of their sexuality (Betten 1993). As Betten (1993:

348) argues: 'The difference in attitude in the various member states toward homosexuals causes a considerable problem for lesbian and gay workers and is an obstacle to their free movement in the internal market.'

Using these terms of reference, in April 1990, MEP Stephen Hughes tabled a question in the EP pointing out that discrimination against lesbians and gay men is an obstacle to the free movement of people and is therefore contrary to the EC's stated aims of creating a free internal employment market and harmonizing national regulations. In the same year, the International Lesbian and Gay Association and the British Stonewall Group also compiled a dossier on the implications of the Single European Market for lesbians and gay men, *Harmonisation within the European Community: The Reality for Lesbians and Gays*, which makes a similar point (Andermahr 1992). But despite clearly demonstrating that the failure to establish equality in all member states for lesbians and gay men runs counter to the three main objectives of the EC, lesbian and gay campaigners once again drew a blank response from the European legislature. The Commission argued that it can act against discrimination only on the basis of nationality and was dismissive of the claim that individual states' discriminatory laws against lesbians and gay men are a barrier to a free internal employment market, arguing that such laws could have only a marginal impact on the free movements of workers (Betten 1993). Rather, the Commission argued it was up to individual member states to outlaw employment discrimination.

Whilst lesbian and gay Euro campaigners continue to join sexual demands to social and economic ones, there is always a danger that if Europe's gay citizens begin to take advantage of more favourable opportunities in liberal member states, then these states may opt or be pushed into falling in line with those that adopt a more restrictive approach towards sexual dissidents. For example, the Netherlands cut back its social security benefits just before the introduction of the internal market for fear of becoming a honeypot for workers from states with less generous welfare policies.

Despite the obvious legal benefits emigration offers to sexual dissidents living in the more restrictive states of the EU, however, the most liberal states are unlikely to be swamped by mass lesbian immigration. The evidence of the UK and Italy suggests that a hostile line by the state towards its lesbian and gay citizens often acts as a catalyst for the development of queer politics and lesbian and gay commercial 'scenes' (Binnie 1994). For example, the introduction by the British Government of section 28 of the Local Government Act 1988 banning the promotion of homosexuality provoked many apolitical or closeted lesbian and gay men into 'coming out' to fight the law. And perhaps most ironically, the struggle to overturn section 28 provided a focus for disparate groups of sexual dissidents, fragmented by geography and politics, to channel their collective social and political energies. Thus Jon Binnie argues that section 28 paradoxically contributed towards making the UK (particularly London), one of the most repressive places in Europe for sexual dissidents, into one of the most lively lesbian and gay scenes on the continent and a popular destination for both tourists and migrants. Many lesbians may therefore be reluctant to swap the social and cultural advantages of living in some cities for the legal benefits of living in more tolerant states.

SPECIFICITIES OF OPPRESSION: MAXIMIZING AND MINIMIZING A EUROPEAN LESBIAN IDENTITY AND POLITICS

Discrimination in the workplace is not only a product of an absence of legal protection, it also depends on cultural pressures that can silence lesbians even when the law is tolerant. In particular, attitudes and ideologies that are situated outside the workplace – for example, in the home, 'community', church and so on – impinge on workplace practices, culture and relationships. Similarly, the ability of lesbians to challenge heterosexism and to 'come out' at work is mediated by the amount of support available from gay political groups and social networks. Both societal pressures and the cultural visibility of lesbians vary in time and space across the EU.

By emphasizing shared or common experiences of discrimination, lesbian and gay transnational political networks often ignore the different attitudes of heterosexual society towards lesbians and the different levels of development that lesbian culture and politics have reached both between and within European countries. Even the 'lesbian look'[7] is not universal, as infamous lesbian photographer Della Grace[8] found when she exhibited her work in Italy. Grace commented in the lesbian magazine *Diva* that

> [s]exual identities based around butch/femme eroticism and sexual style do not appear to operate in ways which I am familiar with. An Italian dyke's greatest fear seems to be being thought of as a man, or too masculine. Whereas the 'super-bitch' is often venerated and ennobled in Northern European and Northern American lesbian circles, in Italy a girl who looks too much like a boy gets a fair amount of criticism and censure from her peers.
>
> (Grace 1994: 31)

In Italy the twin pillars of the Roman Catholic Church and the family have a profound influence on lesbians. In 1992 the Vatican published a document stating that gays and lesbians suffer from an objective disorder and that it is *not* wrong to discriminate against them (*Guardian* 1992). The list of areas in which the Vatican claims it is permissible to discriminate on the grounds of sexuality include the employment of teachers, athletic coaches and the military. Not surprisingly Grace found that there is a reluctance to be 'out' amongst Italian lesbians. She says:

> Italian dykes and faggots have only recently begun to construct a movement around a visible sexual identity. In an odd way, they have Berlusconi and the new fascist coalition government to thank for this, who work like a 'chilli pepper up the ass'.
>
> (Grace 1994: 28)

One such chilli pepper has come in the form of Gianfranco Fini, head of the National Alliance Party, who recently demanded that lesbians and gay men be put to death in concentration camps (*Pink Paper* 1994b).

Grace also identified strong regional differences both in attitudes to lesbians and in lesbian culture and politics. She claims that lesbians in Turin are more closeted and conservative, being reluctant to 'come out' in the workplace; in contrast she argues that Bologna is 'the radical core for queer life in Italy with its large student population and communist city government' (Grace 1994: 28).

In Greece lesbianism is defined as 'unnatural debauchery' and so, by essentialist logic, because lesbian sexuality is 'pathological', lesbians cannot be subject to legal sanction, although there is a subclause in the law that allows a sentence of up to two years' imprisonment to be imposed on women who have sex with underage partners (Rappi 1994). Rappi argues that in Greece, 'deeply influenced by Orthodox Christianity, people can regard homosexuality in men as "weakness", hence forgivable, but lesbianism as "immoral", hence unforgivable'; she quotes a lesbian merchant marine who says: 'Most people think it's better to be a whore than a lesbian' (Rappi 1994: 24). She therefore claims that most women who have sex with other women will not call themselves lesbians because in the absence of naming it is possible to negotiate relationships without risking bringing shame on their families. A Greek woman writing in the lesbian magazine *Diva* explains that

> Greek women will not use the L word to describe themselves, because 'lesbian' has such a denigratory currency, serving to mainly stigmatize older 'predatory' butch women who fail to observe the safe margin of femininity. Since aberrant behaviour produces *Ntropi* [shaming] upon the family name, active sexuality may never be an option for some women.
>
> (*Diva* 1994c: 5)

Although a gay bar culture has emerged in Greece, it has not been matched by the development of a corresponding political culture. But with the greater integration of Europe and a recent higher profile for lesbians on Greek television, Rappi (1994: 26) questions whether 'Greek lesbians [will] now develop a distinct lesbian identity on the western model or whether the situation continues to be one where they remain integrated in the dominant culture, "being it and looking it" but not "declaring it", where blood ties prove to be stronger than homophobia and "the bar" rather than politics is the main axis' (Rappi 1994: 26).

The proposed expansion of the EU to embrace some of the former communist states of eastern Europe looks likely to further complicate attempts to mobilize European lesbians to fight for anti-lesbian discrimination laws and the harmonization of lesbian rights across the EU. Lesbians who grew up under communism have a different history and culture and hence their political priorities differ from those of their western sisters. In communist Czechoslovakia homosexuality was officially 'denied' and open lesbians risked attracting the attention of the secret police. An anonymous Czech lesbian interviewed in the lesbian magazine *Diva* says:

> Gay liberation movements couldn't take place under communism. Lesbianism was not seen in terms of sexuality so much as of gender. So people assumed that if you desired women then you must really want to be a man. Offering lesbians a sex change operation was not official state policy, it simply happened as a result of general attitudes because thinking on these issues was so cut off from an analysis of sexuality.
>
> (Thynne 1994: 22)

Despite the emergence of a more liberal state, the Czech Republic, following the 'Velvet Revolution', she claims that Czech lesbians remain very sceptical of any kind of political movement and have such an aversion to being organized that it will be difficult to mobilize them to fight for anti-discrimination legislation or other political changes. Although the possibility of future EU membership offers Czech lesbians potential social gains, they are critical of their European sisters for trying to impose a western model of 'a lesbian life-style' on them and for not understanding the reluctance of East European lesbians to embrace queer politics (Thynne 1994).

As these very partial snapshots of lesbian culture and the attitudes of heterosexual society to lesbians in Italy, Greece and the Czech Republic highlight, sexual dissidents trying to foster transeuropean political campaigns are caught between the two stools of commonality and difference. On the one hand, given that most lesbians depend entirely on paid work for their economic survival, equality in the workplace would appear to be a priority for gay women in all European states. Certainly the evidence of numerous European surveys, such as that of MEP Vera Squarcialupi, is that lesbians across Europe share common experiences of anti-gay discrimination and harassment at work. On the other hand, however, as this final section of the chapter has highlighted, there are also dangers in claiming a universal European lesbian identity in the rush to persuade the European Union to recognize the rights of its sexual dissidents.

First, a focus on a common lack of legal rights under the law can promote an artificial sense of 'sameness' that obscures complex differences in the attitudes of heterosexual society to lesbians and differences in lesbian history, politics and culture, both within and between European states. Second, UK groups have tended to play a prominent role in organizing European lesbian and gay campaigns, yet British lesbian subculture is strongly orientated towards North America rather than southern Europe, where lesbian culture and identity are not as marked. This has led to accusations of 'British imperialism' and claims that some European groups are being dragged into pursuing an Anglo-American equality agenda rather than being allowed to define their own agendas.

In order to avoid privileging 'sameness' over 'difference', it may be useful for European lesbians to adopt Ann Snitow's notion of 'maximizing' and 'minimizing' identity. Sometimes it is useful to maximize a European lesbian identity; for example, by organizing around specific shared needs such as anti-lesbian discrimination in the workplace. Conversely, it is also important to minimize claims for a common European lesbian identity by recognizing specificities of oppression at national, regional and local levels and the ways in which other identities (age, 'race', class, disability and so on) intersect with and transform the experience of what it means to be 'a lesbian' in different contexts. Rather than setting a fixed European lesbian agenda that requires a forced sense of solidarity, this fluid imagining of 'European' sisterhood allows the tide of communion to ebb and flow. Hopefully, through this process Europe will become not only an equal place to work, but also a more fulfilling place to live for all sexual dissidents.

NOTES

1 As Europe has moved towards greater integration over the last twenty years it has changed drastically in terms of membership, aims and name. Throughout this chapter the acronyms EEC, EC, or EU have been used according to the appropriate level of development of Europe at the time referred to in the text.
2 It is important to remember that many sexual dissidents feel marginalized by the 'lesbian and gay community', which is often perceived to be geared towards white, able-bodied gay men and to have an urban bias.
3 There is obviously a danger that a desire for unity or 'sameness' can lead to the suppression of difference and the false homogenization of a member of a group.
4 The information given about anti-discrimination legislation and lesbian rights in different European member states is to the best of my knowledge correct at the time of writing; however, because the law is constantly subject to change in both the EU and member states it may not be correct at the time of reading.
5 It is well established in the literatures of criminology and sociology that it is very difficult to establish 'real' crime figures because definitions of a crime vary across time and space, as do societal standards about what is and is not acceptable to report. In the same way it is equally difficult to ascertain

'actual' levels of discrimination in the workplace. The higher figures of discrimination recorded in the UK may reflect the fact that British lesbian and gay groups, such as Stonewall, have been more active than their European counterparts in identifying harassment as a political problem and monitoring and publicizing experiences of discrimination.

6 In 1990 WHO ruled that from 1993 onwards homosexuality would be removed from its *International Classification of Diseases*, where it was listed as a mental disorder.

7 There is not and never has been only one lesbian look. However, within North American-British lesbian culture there are a number of stereotypical 'lesbian looks' that women use to 'dyke spot', such as 'butch-femme'. As Grace explains in the text these 'stereotypical looks' do not seem to be understood in the same way in Italian lesbian culture.

8 Della Grace is a lesbian photographer 'infamous' for her portraits of s/m dykes.

REFERENCES

Andermahr, S. (1992) 'Subjects or citizens? Lesbians in the new Europe', in A. Ward, J. Gregory and N. Yuval Davis (eds) *Women and Citizenship in Europe: Borders, Rights and Duties: Women's Differing Identities in a Europe of Contested Boundaries*, London: Trentham Books.

Betten, L. (1993) 'Rights in the workplaces', in K. Waaldijk and A. Clapham (eds) *Homosexuality: a European Community Issue: Essays on Lesbian and Gay Rights in European Law and Policy*, Dordrecht: Martinus Nijhoff.

Binnie, J. (1994) 'Invisible Europeans: sexual citizenship in the new Europe', paper presented at the regional conference of the International Geographical Union, Prague, Czech Republic.

Burke, M. (1993) *Coming out of the Blue*, London: Cassell.

Diva (1994a) 'MOD keeps "lesbian index"' (October): 9.

—— (1994b) 'Lesbian Avengers campaign kicks off with series of zaps' (October): 9.

—— (1994c) 'Greek pansexuality is not so queer' (October): 5.

Evans, D. (1993) *Sexual Citizenship: The Material Construction of Sexualities*, London: Routledge.

Fiske, J. (1993) *Power Plays, Power Works*, London: Verso.

Grace, D. (1994) 'Ciao Della!', *Diva* (August): 28–32.

Guardian (1992) 'Gays angry as Vatican backs bias' (24 July): 22.

Gutek, B. (1989) 'Sexuality in the workplace: key issues in social research and organizational practice', in J. Hearn, D. Sheppard, P. Tancred-Sherriff and G. Burrell (eds) *The Sexuality of Organization*, London: Sage.

Hall, M. (1986) 'The lesbian corporate experience', *Journal of Homosexuality* 12, 3/4: 59–75.

—— (1989) 'Private experiences in the public domain: lesbians in organizations', in J. Hearn, D. Sheppard, P. Tancred-Sherriff and G. Burrell (eds) *The Sexuality of Organization*, London: Sage.

Hugill, B. (1992) 'MoD to review forces homosexuality ban', *Observer* (22March): 3.

Jay, K. and Young, A. (1977) *The Gay Report*, New York: Summit.

Jury, L. (1993) 'Gays suffering high levels of harassment', *Guardian* (20 November): 11.

Khayatt, M. D. (1992) *Lesbian Teachers: An Invisible Presence*, New York: State University of New York Press.

Leathwood, C. (1994) book review of M. D. Khayatt's *Lesbian Teachers: An Invisible Presence, Gender & Education* 6, 1: 96–8.

Leidner, R. (1993) *Fast Food, Fast Talk: The Routinisation of Everyday Life*, Berkeley: University of California Press.

Lorenz, B. (1992) 'German lesbians claim marriage rights', *Lesbian London* 9: 2.

McDowell, L. (1995) 'Body work: heterosexual gender performances in city workplaces', in D. Bell and G. Valentine (eds) *Mapping Desire: Geographies of Sexualities*, London: Routledge.

Murray, H. (1994) 'Butch bomberas & femme bolleras', *Diva* (August): 16–18.

Phelan, S. (1989) *Identity Politics: Lesbian Feminism and the Limits of Community*, Philadelphia: Temple University Press.

Pink Paper (1994a) 'Truth will out on police records' and 'BT pressure over MoD lists', 3 (February): 6.

—— (1994b) 'Mortal combat', 3 (February): 14.

Pringle, R. (1988) *Secretaries Talk: Sexuality, Power and Work*, Verso: London.

Rappi, N. (1994) 'Cool in a hot climate', *Diva* (August): 24–8.

Raz, J. (1986) *The Morality of Freedom*, Oxford: Clarendon Press.

Rose, K. (1994) *Diverse Communities: The Evolution of Lesbian and Gay Politics in Ireland*, Cork: Cork University Press.

Simpson, M. (1992) 'A secret war', *Guardian* (March).

Squirrell, G. (1989) 'In passing teachers and sexual orientation', in S. Acker (ed.) *Teachers, Gender and Careers*, London: Falmer Press.

Tanca, A. (1993) 'European citizenship and the rights of lesbians and gay men', in K. Waaldijk and A. Clapham (eds) *Homosexuality: A European Community Issue: Essays on Lesbian and Gay Rights in European Law and Policy*, Dordrecht: Martinus Nijhoff.

Tatchell, P. (1992) *Europe in the Pink: Lesbian and Gay Equality in the New Europe*, London: GMP Publishers.

Taylor, N. (1986) *All in a Day's Work: A Report on Anti-lesbian Discrimination in Employment and Unemployment in London*, London: Lesbian Employment Rights.

Thynne, L. (1994) 'From velvet to sexual revolutions?' *Diva* (August): 2–24.

Valentine, G. (1993) '(Hetero)sexing space: lesbian perceptions and experiences of everyday spaces', *Environment and Planning D: Society and Space* 11: 395–413.

Waaldijk, K. (1993) 'The legal situation in member states', in K. Waaldijk and A. Clapham (eds) *Homosexuality: A European Community Issue: Essays on Lesbian and Gay Rights in European Law and Policy*, Dordrecht: Martinus Nijhoff.

Wilson, A. (1993) 'Which equality?', in J. Bristow and A. Wilson (eds) *Activating Theory: Lesbian, Gay and Bisexual Politics*, London: Lawrence & Wishart.

6

FAMILY POLICIES AND WORKING MOTHERS: A COMPARISON OF FRANCE AND WEST GERMANY

Jeanne Fagnani

Although most member states of the European Union (EU) recognize the importance of the family and of the state's obligations towards the family, few countries, with the exceptions of France and Germany, have created a specific family policy distinct from other aspects of social welfare. Extensive differences exist between the principles and objectives of member states, reflecting the diversity of national priorities. Thus, for example, the relative part of the public budget devoted to family payments, the methods of financing and the qualifying conditions for such funds differ from one country to another.

Even the concept of a 'family policy' is difficult to define and refers to a variety of realities. An explicit and coherent family policy rarely exists. Rather, such a policy often consists of an amalgam of elements of other policies. In reality, all public interventions in the areas of justice, education, health, housing and employment have an impact, directly or indirectly, on the well-being of families and have repercussions on their behaviour and their disposable revenue. Inevitably, the family is involved in a transversal fashion in all social affairs. In all countries, the modalities and characteristics of family policies are tightly linked to other social policies. By 'family policy', one might then include all legal measures taken by public authorities (at whatever local, regional, or national level) which affect, either directly or indirectly, the style of life, the available resources and, in general, the well-being of families. Nevertheless, in most countries, there is usually a patchwork of measures inspired by a wide variety of principles and influences.

Analysis of the literature on social and family policies in Europe (whether explicitly defined or not) shows that a global and unrestrictive approach is required in order to understand the logic supporting them and the philosophies behind them. Only by taking into account all the dimensions of the family policy followed by each country can one really evaluate the interactions between policies and family behaviour, and in particular women's behaviour, in different areas such as employment, child-bearing and the education of children.

Furthermore, while most member states of the European Union recognize explicitly in their constitutions (except for the UK which does not have a written constitution) the importance of the family and the obligations of the state to it (except for Belgium, Denmark and the Netherlands), only Germany, France and Luxemburg have a minister or secretary of state explicitly charged with the family portfolio, confirming, on an institutional level, the importance accorded to this subject. Among these three countries, France stands out by the extent of its legislation and the benefits it provides (both in cash and in kind, and especially in the public provision of childcare) in support of mothers who wish to or must work outside the home. Does this explain why, although the birth rate is higher in France than in West Germany, French mothers are often more active economically than their West German counterparts? Can a family policy have an impact on fertility rates if policy-makers are reluctant (as in Germany), for historical and cultural reasons, to help parents, and especially mothers, combine professional and family life?

A comparative analysis between France and West Germany can help to clarify some of these issues, since the two countries represent extremes in approaches to public policy towards working mothers (Fagnani 1992). I will therefore focus on the interaction between public policy for the family, the wider cultural context, professional attitudes and working patterns for women in France and West Germany. I selected West Germany, rather than Germany as a whole, because the former GDR had a completely different welfare regime (Merkel 1992) and because the German reunification is too recent to examine the effects of policy changes in East Germany on working mothers and fertility levels. Such an analysis would require a separate study. Moreover, as Ostner (1993: 93) stated, 'East German women are being fitted more or less willingly into the West German regime' and the East German child- and mother-centred policies were or soon will be abolished by the transformation process. The present analysis is limited to the national level, and hence does not raise questions about differences at other scales, for example, of variations across regions within the respective countries, or by class or ethnicity. These questions might, nevertheless, be explored in later work.

This chapter is based on the hypothesis that the extent of the different measures implemented to help couples combine their family obligations with their professional lives influences the range of possibilities and strategies that women, in particular, try to develop as mothers and paid workers. None the less, the extent and the level of assistance offered to families not only determine the objective possibilities available to mothers in trying to satisfy their aspirations, they also help to legitimate those aspirations for the mothers themselves. Accordingly,

guilt-inducing, environment for mothers who want to be economically active. Nevertheless, these policies are the expression of dominant value systems. It could be argued that the content and nature of the different policies drawn up for families, on the one hand, and women's expectations, on the other, are shaped in such a way as to form a coherent interactive system. While, in West Germany, for example, family policy confirms and strengthens the antagonism between maternity and employment (Schultheis 1987), in France the model of the working mother (following the Swedish and Danish examples) is fully integrated into family policy. Since the interwar period, successive French governments have developed concerted social policies designed to enable women to engage in full-time employment while also raising a family (Messu 1992).

CHILD-BEARING IN FRANCE AND WEST GERMANY

As in the other EU countries, birth rates have decreased in both France and West Germany since 1964. Between 1965 and 1992, the total period fertility rate declined from 2.84 per 1,000 to 1.73 in West Germany and from 2.51 per 1,000 to 1.40 in France. In both countries, the completed fertility rate had began to decline long before the Second World War. However, for women born in 1955, the completed fertility rate is still much higher in France than in West Germany: 2.12 compared with only 1.66 children (and 2.64 and 2.17 children respectively for women born in 1930) (Eurostat 1993). Germany also has a higher proportion of couples with only one child, while families with three or more children are quite rare (Bertram 1991; Süttzer and Schwartz 1990). For example, in 1991, 20 per cent of West German women born in 1955 remained childless, compared with only 9.0 per cent of French women, while 17 per cent of these German women have three or more children, compared with 32 per cent of French women (Figure 6.1). In 1991, 15 per cent of West German women aged 35–9 living in private households have no children, 27 per cent have only one child and 58 per cent have two or more children compared with 10 per cent, 23 per cent and 67 per cent respectively in France (Eurostat 1993).

As many studies have shown, no direct causal relationship can be established between the decline in the birth rate and women's increasing labour force participation. Although the fertility level is higher in France than in West Germany, French mothers are more often economically active than German mothers.

FRENCH MOTHERS MORE ECONOMICALLY ACTIVE THAN GERMAN MOTHERS

Female participation in paid employment has increased significantly in recent decades throughout Europe. As far as France and West Germany

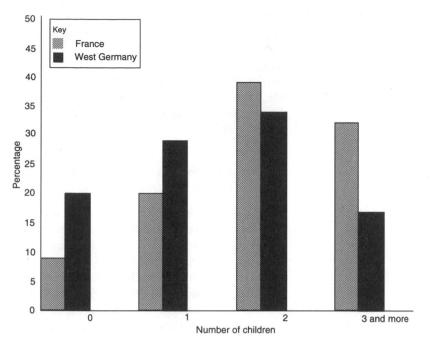

Figure 6.1 Women born in 1955 according to number of children

are concerned, the expansion of the female workforce has taken place almost exclusively through an increasing utilization of the labour potential represented by married women. In 1991, the female labour force participation rate in France was one of the highest in the EU (44.3 per cent), together with Denmark (46.6 per cent) and the United Kingdom (43.2 per cent), compared with 42.5 per cent in Germany (Eurostat 1993). The activity rate for women aged 25–49 was also higher in France (74 per cent) than in West Germany (67 per cent).

In France, there is a long tradition of female participation in the workforce: just before the First World War approximately 20 per cent of women worked outside the home, while only from 13.5 per cent to 15.5 per cent of women were active in Germany, England and Italy (Marchand and Thélot 1991). In 1970, the employment rate for women aged 20–64 was 55 per cent in France compared with 42 per cent in West Germany. In 1990, these differences in female employment patterns persisted between the two countries. For example, 75 per cent of all French women 30–4 years old were economically active (compared with 65 per cent in West Germany) and their working behaviour was drawing ever closer to the male model. In France, it has become the norm to be a working mother with two children, even while they are babies. The most qualified women are the most inclined to work and they continue working when they have children: 85 per cent of highly qualified mothers with two children are employed as opposed to 57 per cent of mothers without higher level diplomas (Desplanques 1993).

In both countries, child-bearing and motherhood have a clear impact on the activity rate of women (Desplanques 1993; Ostner 1993). However, French mothers are more often economically active than their German counterparts, irrespective of the number of children and the age of the youngest child (see Table 6.1).

French mothers do less part-time work than German mothers

In both France and the former Federal Republic of Germany, a variety of legislative and fiscal measures has been introduced to encourage the development of part-time employment as a way of helping mothers reconcile their professional and family lives. Part-time work has become more widespread in a context where more flexible worktime practices, such as temporary and casual employment, have arisen due to the reorganization of production and changes in work organization. In 1991, the percentage of women in part-time employment was 30.1 per cent in Germany and 23.5 per cent in France.

In France, as in Germany, the so-called women-friendly initiatives, such as life career pauses, job-sharing and flexible working patterns, may make working mothers' lives easier. However, these measures may actually reduce the number of women who choose to fight their way to the top of the career ladder. In France, a recent study (Maruani and Nicole 1989) has demonstrated that, for the majority of working women, flexibility at work has meant poor job security. Most of the women who

Table 6.1 Activity rate for women aged 20–59 by age of youngest child and by number of children, 1990

	West Germany	France
Number of children		
No children	67.3	68.6
1 child	59.6	77.3
2 children	48.6	69.9
3 or more children	36.0	39.6
Age of youngest child		
2 years or less	41.5	61.5
Over 2, under 7	52.8	69.4
Over 7, under 14	61.5	72.7
Youngest under 7	46.7	65.1
Youngest under 14	49.8	66.4

Note Activity rates for mothers relate to women who are married to the household head. Data for other mothers, such as living with their parents, are unavailable.
Source Eurostat 1990; data from Community Labour Force Survey

work part-time are underqualified and badly paid. This phenomenon maintains and reinforces the gender discriminations in the labour market.

Reliance upon flexible types of employment is, however, much more frequent among German mothers than among their French counterparts (for example, 43.0 per cent among German women aged 20–39 who work part-time have at least one child under 5, compared with 28.0 per cent of such mothers in France in 1991) (Eurostat 1993).[1] Among economically active women aged 25–44 years old who have one or two children, 59.5 per cent in Germany, compared with only 25.7 per cent in France, combine employment and caregiving by working part-time. Among women with three or more children, the proportion is 50.4 per cent and 36.8 per cent respectively for the two countries (Kempeneers and Lelièvre 1991). Working patterns are less affected by motherhood and child-raising in France than in West Germany. French mothers have more continuous employment patterns than do German mothers (Kempeneers and Lelièvre 1991). For example, during the pre-school period of the first child, 64 per cent of German women stopped work compared with only 41 per cent of French women, and 18 per cent and 47 per cent, respectively, worked full-time.

Comparisons of women in the educational systems in the two countries show that French women have established a stronger position in higher education. Between 1985 and 1990, the increase in student numbers was significantly higher for women than for men in all member states other than in West Germany (Eurostat 1993). At the age of 18, over 86 per cent of girls are still at school in France but only 75 per cent in West Germany; by the age of 19–20, about 60 per cent of girls are still at school or university in France compared with 45 per cent in Germany (Debizet 1990). In 1991, the ratio of women to men in the student population was 1.12:1 in France compared with 0.67:1 in West Germany (Eurostat 1993). Subsequently, French women demonstrate a greater attachment to a career by more often maintaining an uninterrupted work history. These findings suggest that French women in recent generations have adopted innovative attitudes towards work more rapidly than West German women, thereby breaking away from the patterns of behaviour followed by their mothers.

THE DIFFERENCES BETWEEN FRANCE AND WEST GERMANY: A COMPLEX BUNDLE OF EXPLANATORY FACTORS

In trying to explain these differences, it is important to take account of both cultural contexts and family policies, and especially of measures designed to help women reconcile their family and work roles.

Value systems and cultural differences

The attitudes and behaviour of women with regard to employment and the family are strongly linked to value systems which are specific to each couple. They underpin strategies developed within couples. The findings from opinion surveys provide some illuminating insights into changing attitudes on these issues. A survey of political and social values conducted in 1991 by the Times Mirror Center for People and the Press (1991) in France, West Germany and other European countries revealed that sharp cultural differences persist between these countries.

A distinction was found between the French and German respondents when Europeans were asked whether they preferred a family situation where the husband worked while the wife stayed at home to look after the house and children, or a family situation where both had jobs and shared responsibility for the home and children. In France, 64 per cent of respondents opted for the both-at-work scenario, compared with only 54 per cent of West Germans. In another survey conducted in 1987 by Eurobaromètre (1990), respondents were asked which of the following three arrangements corresponded most closely to their idea of a family: a family in which both husband and wife have equally absorbing work and in which household tasks and childcare are shared equally between husband and wife (egalitarian option); a family in which the wife's work is less absorbing than the husband's, and in which she takes on more of the household tasks and childcare (medium option); or a family in which only the husband works and the wife runs the home (homemaker/breadwinner option). In France only 22 per cent of the women surveyed supported the traditional option (homemaker/breadwinner option) compared with 34 per cent of German women and 29 per cent of German men. In France 48 per cent of the women but only 27 per cent of German women and 25 per cent of German men supported an egalitarian division of labour. In West Germany, the proportion of men and women preferring the traditional homemaker/ breadwinner model is greater than the proportion preferring the equal roles model.

Considerable social pressure is still exerted on mothers in West Germany, in comparison with France, to devote themselves exclusively to bringing up their children during the early years (Bertram 1991; Ostner 1993). Public opinion readily identifies the image of the mother as superwoman with that of a bad mother. The *Rabenmutter* or mother-crow is the mother who allows somebody else to look after her children. In order to overcome these pressures and remain in the labour market, a mother with a child aged under 3 must demonstrate single-minded determination. German public opinion upholds the theory that a young child needs its mother to be with it at all times and that any separation is traumatic for the child. Both the Catholic and Protestant churches, which continue to play an important role in the social and economic life of the country, contribute to this process of loading guilt on to working mothers, as do numerous educators and paediatricians.

In France, by contrast, the early socialization of young children is not only accepted but is also socially valued. As demonstrated in previous research (Fagnani and Meunier 1992), well educated women, for example, feel that they must rise to the challenge by excelling not only in their employment but also in their family roles, by demonstrating their ability to manage the two areas of their lives without sacrificing one to the other. At the same time, women in France have been supported and encouraged by the state in their efforts to combine family and professional life. To some extent, state intervention can be construed as recognition of their contribution to the community both as mothers and as workers.

Normative models of the family

West German family and social policies can be characterized as a strong male breadwinner welfare regime (Lewis 1993; Ostner 1993; Schultheis 1993). In comparison with Germany, the French social welfare system continues to be characterized by its concern with the family and, more specifically, with the importance attributed to demographic objectives. Over the years, successive French governments have put in place an impressive array of measures designed to encourage family building and to create an environment in which the welfare of children is recognized as paramount and is overtly supported by the state (Hantrais 1992).

Demographic issues have been an important underlying concern in family policy in France. Policy-makers have understood, however, that policies discouraging women from working would not necessarily raise the birth rate. They are also aware that financial incentives in the form of child allowances or paid maternity leave are not sufficient to encourage women to have more children without adversely affecting their economic activity. French policy-makers have made the explicit assumption that women want to be able to work while also raising children. As a result, France is at the forefront of developments in employment policy designed to make child-rearing compatible with employment. One policy objective has thus been to provide choice by making a range of options feasible. In West Germany, by contrast, the assumption is that women should and do want to stay at home to look after young children. Therefore, within the European context, France and West Germany offer examples of the two ends of the spectrum of childcare provision, which is one of the most important measures designed to reconcile family and professional life.

France is amongst the EU countries (as are Belgium and Denmark) with the most generous provision for childcare. Immediately following the Second World War, the French government set up community-funded day-care centres (crèches) in an attempt to attract women into the workforce at a time of acute labour shortage (Norvez 1990). The

1980s were characterized by measures intended to help couples combine family and professional life; for example, through childcare allowances, tax relief and parental leave. By contrast, in West Germany, there is a severe shortage of childcare facilities. Although there are more private facilities, only 3 per cent of children under 3 are cared for in public day-care centres, and fewer than 40 per cent of 3-year-olds attend a nursery school, compared with 20 per cent and more than 90 per cent respectively in France (Melhuish and Moss 1991). German kindergartens, which are not part of the educational system (60 per cent are managed by the churches), are often open only in the morning. Of children aged 3 to 6 years old, 60 per cent are cared for by local government services in Germany (for four to six hours a day) compared with 95 per cent in France (for eight hours a day). In France, extensive public provision has also been made for care before and after school hours, and an important network has developed throughout the country of registered childcare workers, closely supervised by local authorities. Parents who decide to set up their own *crèche parentale* (family playgroup) are also eligible to receive subsidies.

Recent legislation in France making it possible to deduct a higher proportion of child-minding expenses from taxable income has provided a financial incentive for parents to engage childcare services. An allowance (subsidy for childcare in the home) of 2,000 francs per month is also given to working mothers towards the cost of having a child looked after in its own home by an approved child-minder (*assistante maternelle agréee par la DDASS*) until the child reaches the age of 3. The allowance is intended to cover the child-minder's social insurance contributions.

As far as parental leave and career breaks are concerned, France is one of the EU countries with the most generous arrangements. Both men and women are legally entitled to take parental leave for three years for each child at the end of maternity leave on the condition that the employee concerned has been with an employer for at least a year. The period of leave can alternate between parents. Since 1986, as part of government policy to encourage families to have a third child, parents with three or more children have been paid flat-rate benefits (not income-related and not taxable) during parental leave (at 2,871 francs a month in 1993). Parents are eligible for this paid parental leave (APE) only if they have worked for at least two years in the ten years preceding the birth. After parental leave, an employee must be reinstated without reduction in pay in the same position or in a similar one, and is eligible for retraining with pay.

In West Germany, by contrast, paid parental leave (*Erziehungsgeld*, a benefit created in 1985) is granted even for the first child (600 DM per month in 1993), and it is not necessary to have worked previously, unlike in France. Leave with pay is granted for two years, but after the child reaches the age of 7 months certain restrictions are imposed: the benefit becomes means-tested. The concept behind paid parental leave in

Germany is different from that of the APE in France. In Germany, leave is considered as a maternal salary (although fathers may also request it)[2] granted to mothers even if they have never worked, while in France leave is offered to enable mothers with three children or more to take time off from professional life.

In 1991, in reunified Germany, 790,000 women received this benefit (which accounted for 95 per cent of all the births which occurred that year and represented an expenditure, for the Federal government, of 6 billion DM or about US$3,500 million). Landenberger (1991) argued that the generous parental leave programme (financed by the old age insurance system) works at the expense of policies to extend public childcare and has, at best, served to make women's paid and unpaid work more flexible. In France, during the same period, there were only 171,000 recipients (99 per cent of which were women), about 40 per cent of the eligible persons, representing an expenditure of about 5.7 billion French francs (about US$950 million) essentially financed through employers' contributions and paid by the social security system.

CONCLUSION

The role played by the state in shaping the family–employment relationship should not be underestimated. Under the French family policy, priority is placed upon enabling mothers to remain in the workforce. But policy and legislation also have a symbolic component: French mothers feel that their economic activity is recognized as being legitimate. The women's rights movement would appear to have had an impact on attitudes at an official level, as testified by the number of ministerial appointments in the area of women's rights and welfare and family policy.

It can be argued that family size has been maintained in France in conjunction with full-time employment for women because they have material and psychological support from society and its institutions. French women may leave the labour market if they are planning to have a large family and they may choose to devote themselves to a family career, but they are not being forced to stop because they cannot manage to combine work and a family (Fagnani 1992; Hantrais 1992).

In West Germany, by contrast, where social policy has not been conceived with the object of providing the conditions needed so that women can combine employment with family life, the most obvious choice for the majority of couples is either for both husband and wife to work full-time and not to have children (or to limit the number of children to one), or, if they decide to have children, for the woman to stop working temporarily or change to part-time hours at least until the children are old enough to manage by themselves. The German example suggests that by discouraging mothers from working, public authorities participate in the maintenance of the low German birth rate. The ideological conceptions underlying actions in this area would, however, seem to correspond more or less to the dominant principles in German public opinion concerning the role of mothers. Analysis of the interactions between government policies and women's working and family lives does shed some light on both the behaviour of mothers and the impact of government policies on families.

NOTES

1 Unlike the national surveys which tend to set minimum and/or maximum hours worked, the EC *Labour Force Surveys* use no such definition, although a distinction is made between occasional (seasonal) and principal (regular) employment. In these biennial surveys, the respondent is expected to use his/her own judgement on whether he/she is employed part-time or full-time. In France and in West Germany, part-time women workers are heavily concentrated in the service sector (respectively 82.1 per cent and 77.1 per cent) (Eurostat 1989).

2 Although policy statements in France and in Germany stress that the opportunity to take parental or family leave, to work part-time or to job-share is available to both men and women, it is extremely rare for a man to become a househusband or for him to request time off to care for sick children. Men (and often also their wives) are reluctant to make such a request, not only as a result of social pressures, but because they do not want family problems to interfere with their work commitments.

REFERENCES

Bertram, H. (ed.) (1991) *Die Familie in Westdeutschland: Stabilität und Wandel familialer Lebensformen*, Opladen: Leske-Budrich.

Debizet, J. (1990) 'La scolarité après 16 ans', in *Données Sociales*, Paris: Institut National de la Statistique et des Etudes Economiques (INSEE).

Desplanques, G. (1993) 'Activité féminine et vie familiale', *Economie et Statistique* 261:23–32.

Eurobaromètre (1990) *L'opinion publique dans la Communauté européenne*, Brussels: Commission des Communautés Européennes.

Eurostat (1989) *Labour Force Survey*, Luxemburg.

—— (1990) 'Network of experts on the situation of women in the labour market', *Bulletin on Women and Employment in the European Community*, no.1, Luxemburg.

—— (1991) *Données démographique européennes*, Luxemburg.

—— (1993) *Rapid reports: Population and Social Conditions*, no. 10, Luxemburg: Office for Official Publications of the European Communities.

Fagnani, J. (1992) 'Fécondité, travail des femmes et politiques familiales en France et en Allemagne de l'Ouest: les Françaises font-elles des prouesses?', *Revue Française des Affaires Sociales* 2: 129–47.

Fagnani, J. and Meunier, C. (1992) 'Modèles culturels, interactions conjugales et fécondité', *Recherches Sociologiques* 23, 1: 123–42.

Hantrais, L. (1992) 'La fécondité en France et au Royaume-Uni: les effets possibles de la politique familiale', *Population* 4: 987–1016.

Kempeneers, M. and Lelièvre, E. (1991) *Emploi et Famille dans l'Europe des Douze*, Brussels: Commission des Communautés européennes, Eurobaromètre 34.

Landenberger, M. (1991) *Die Beschäftigungsverantwortung der Rentenversicherung*, Berlin: Edition Sigma.
Lewis, J. (ed.) (1993) *Women and Social Policies in Europe*, Aldershot: Edward Elgar.
Marchand, O. and Thélot, C. (1991) *Deux siècles de travail en France*, Paris: INSEE.
Maruani, M. and Nicole, C. (1989) *La flexibilité à temps partiel*, Paris: La Documentation Française.
Melhuish, E. C. and Moss, P. (1991) *Day Care for Young Children: International Perspectives*, London/New York: Routledge.
Merkel, I. (1992) 'Another kind of women', *German Politics and Society* 24/25: 1–9.
Messu, M. (1992) *Les politiques familiales*, Paris: Les Editions Ouvrières.
Network of Experts on the Situation of Women in the Labour Market (1992) *Bulletin on Women and Employment in the EC*, Brussels: Commission des Communautés européennes.
Norvez, A. (1990) *De la naissance à l'école. Santé, modes de garde et préscolarité dans la France contemporaine*, Paris: PUF-INED.
Ostner, I. (1993) 'Slow motion: women, work and the family in Germany', in J. Lewis (ed.) *Women and Social Policies in Europe*, Aldershot: Edward Elgar.
Schultheis, F. (1987) 'Fatale Strategien und ungeplante Konsequenzen beim Aushandeln Familiarer Risiken zwischen Mutter, Kind und Vater Staat', *Sociale Welt* 38: 34–57.
—— (1993) *La famille: une catégorie de droit social? Analyse comparative de la prise en compte des situations familiales dans les systèmes de protection sociale en France et en Allemagne*, Paris: Caisse Nationale des Allocations Familiales.
Stutzer, E. and Schwartz, W. (1990) 'Childbirth and female employment in the Federal Republic of Germany', in *Study on the Relationship between Female Activity and Fertility*, Eurostat, Brussels: Commission des Communautés européennes.
Times Mirror Center for People and the Press (1991) *The Pulse of Europe: A Survey of Political and Social Values and Attitudes*, Washington, DC: Times Mirror Center.

7

AT THE CENTRE ON THE PERIPHERY? WOMEN IN THE PORTUGUESE LABOUR MARKET

Isabel Margarida André

INTRODUCTION

Women's employment in Portugal presents some features that are difficult to understand through the traditional analytical models developed in different socioeconomic contexts – those of the 'centre'. Portuguese women's rate of economic activity is actually one of the highest in the European Union (EU), part-time jobs are unusual, women's access to central segments of the labour market is relatively large and, among young people, women show a high level of education. At the same time, women are clearly marginal with respect to occupational segmentation and are the large majority of the long-term unemployed. These characteristics should be understood not only in terms of employment demand but also with respect to the particularities of social regulation over recent decades: the Colonial War (1961–74), the emigration process (in the 1960s), the revolution (1974) and the entry into the European Community (1986) have defined important conditions for women's participation in the labour market. Further, rapid changes in social and cultural values and representations, namely in family organization, including the strong decline in fertility, cannot be dissociated from the changes in the labour market. It is important to highlight this relation between public and domestic spheres in order to understand the particular characteristics of women's employment in Portugal in the 1990s. To provide a framework for interpreting the Portuguese experience, I will begin by examining the intersections between gender relations, capital and various social agents such as the state, employers and trade unions, and then elaborate on the context of women's integration into the labour market.

CONVERGENCES AND CONFLICTS BETWEEN CAPITALISM AND PATRIARCHY: PARTICULAR CHARACTERISTICS IN SEMI-PERIPHERAL AREAS

In many ways, the specific nature of jobs reserved for women in Portugal is typical of other semi-peripheral areas and their dominant model of development. In order to understand the gender segmentation of employment, however, it is necessary to consider the underlying processes shaping the relationship between a basically Fordist, state-made social order and labour relations in a situation where two economic sectors coexist – a modern industrial sector in the process of attempting to make the transition from a Fordist to a flexible mode of production and a large traditional sector based on extensive accumulation. To interpret the marginalization of women's labour in this context we have to take into account the interpenetration of the only apparently separate public and private realms (see Figure 7.1).

The public sphere

The roles performed by the various protagonists in social regulation, their alliances and their battles, have become increasingly more decisive in mapping out the labour market, especially its gender segmentation, whether this occurs through formal channels of laws and regulations or through the influence exerted on behaviour and social representation (Walby 1986; Walby 1990). In the last decade, the state, the trade unions and the entrepreneurs' associations have become less active players while professional associations, consumer associations, ecological movements, or feminist movements have commonly emerged in more developed countries as new regulating agents. In Portugal, this transition takes on special significance. On the one hand, trade union organizations and entrepreneurial associations, as well as the bargaining procedures that are followed between the two, have a short history, dating back only to 1974, and thus have not had enough time to become deeply entrenched (Sousa Santos 1993). On the other hand, the weaknesses and incoherencies of the Fordist order may well result in the appearance of less rigid forms of regulation and give rise to non-traditional actors who exert influence mainly over behaviour and social representation. The gender segmentation of the labour market, however, is not only due to the processes of social order, it also has its origins in the characteristics of the workers and in entrepreneurial strategies. People's geographical mobility, their availability and their educational levels are vital conditioning factors in segmenting the labour force. Internationally, over the last few decades, entrepreneurs have promoted job flexibility as a common management strategy, leading to a redefinition of the segments making up the labour market (Lipietz and Leborgne 1992). These strategies (including functional flexibility, mainly based on skills, polyvalence and professional

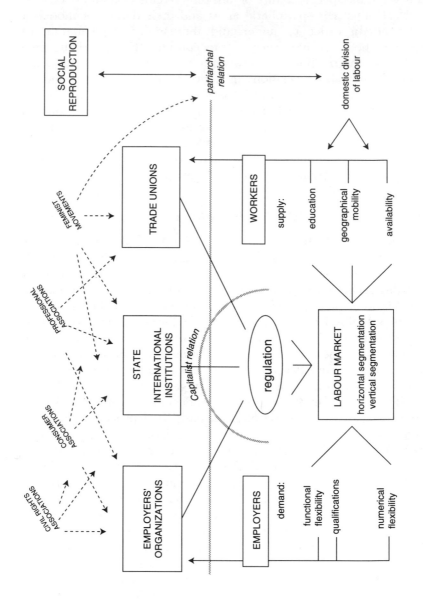

Figure 7.1 Gender segmentation of the labour market: processes and protagonists

mobility and numerical flexibility which, above all, implies rapid changes in the volume of jobs and the different ways of staggering timetables) have made a decisive impact on job stability and have seriously affected women's employment. In Portugal, because a large sector of the economy continues to be based on the extensive accumulation system that benefits capital mainly by preserving low production costs, strategies of numerical flexibility predominate. If, on the one hand, women have been more severely hit by generalized job instability, on the other hand, gender segmentation caused by demands for greater availability and mobility – which are the features of functional flexibility – has not become as widespread in Portugal as it has elsewhere.

The private sphere

In Portugal, as in the other southern European countries, semi-peripheral conditions and the associated late arrival of industrialization and urbanization have been important in perpetuating traditional family ties. As a result, many more women have been able to draw on their families (especially their families of origin) for help while pursuing their professional activities than have women in countries where the immediate, nuclear family is the common form. In addition, slow progress in providing outside services to meet domestic responsibilities (mainly childcare and care for the sick and the elderly) acts as a brake on women's professional ambitions and, in particular, severely limits their mobility.

Patriarchal relations in the public sphere

Even though Portuguese law guarantees formal equality between men and women, the state has never refrained from transmitting patriarchal models. This circumstance has affected numerous legal cases, mainly those concerned with sexual harassment or the definition of maternity and paternity (Beleza 1990), which have all been allowed to slip through the jurisprudence net. Inequality in terms of treatment has also materialized in clauses contained in countless instruments pertaining to collective labour (ratified by the Ministry of Labour, by trade unions and by entrepreneurs' associations). Such regulations frequently and blatantly infringe upon anti-discriminatory principles – the grammatical gender (i.e. masculine inflections) of the various professional categories mentioned and the definition of rights inherent to parenthood and assistance to the family are the most common infractions. The lack of monitoring and the difficulty in getting through to the lawcourts clearly indicate that the patriarchal models prevalent in the various institutions that shape social matters will continue to hold sway.

THE CONTEXT OF WOMEN'S INTEGRATION
INTO THE LABOUR MARKET IN PORTUGAL

In addition to Portugal's being conditioned by its semi-peripheral position on the fringe of the European Union, the integration of Portuguese
women into its labour market over the past few decades has been
indelibly marked by specific local influences (Silva 1993; André 1991;
André 1994). Up to the 1960s the importance of farming in the
Portuguese economy and the influence of Catholicism in Portuguese
society were especially significant. Until that time, agriculture employed
most of the active population although its nature varied greatly between
different regions of the country. In the north, small family holdings
predominated with few marketable products. In these rural communities women played an important social role that was intrinsically
associated with the lack of distinction between production and reproduction. Chores within the family unit were of a complementary nature and
it was not possible to consider some more important than others. In the
south, the system of large landed estates was responsible for women's
early absorption into the labour market where they occupied a place on
the fringe, doing all the most unrewarding jobs at extremely low wages
and more frequently than not on a temporary basis.

The overriding influence of Catholicism on Portuguese society,
together with the conservative ideology of the Salazar regime (the
alliance was given legal status in 1940 when an official agreement was
drawn up between the state and the Church), was responsible for
diffusing a model of feminine behaviour that stifled any move to
emancipate women for decades. This model restricted woman's functions to the roles of mother and wife, condemning women's participation
in all walks of public life and ennobling their submissive attitudes. The
effects of this ideology have endured up to the present and have imposed
a severe handicap on women's professional fulfilment.

From the beginning of the 1960s, the situation began to change,
primarily as a result of the industrial impetus, the migratory process
and the Colonial War (Silva *et al.* 1984; Ferrão 1985). Industrialization,
which was partially related to incoming foreign capital that became
available when Portugal entered the European Free Trade Association
(EFTA) in 1959, meant that labour was densely concentrated into the
two main urban-industrial areas of Lisbon and Oporto; this development
set in motion large-scale migration within the country. The process gave
rise to important changes in family organization – the traditional
extended family being replaced by the immediate nuclear family. At
the same time, intensive suburbanization took place with all the unfavourable consequences for women's daily life that had occurred earlier in
many other European countries; isolation and removal from their
families of origin were the most negative effects of this process.

In the mid-1960s, industrial investment, which had hitherto caused
growth in employment only for men, started to direct its preference
towards sectors employing more women, such as the clothing industry,

electrical goods assembly and the food industry. The abundance of a cheap labour force was a basic criterion for setting up these new firms. They concentrated business in the two metropolitan areas where the large number of newly arrived, available women acted as an important incentive.

These new industries did not provide the main thrust behind the growth in women's professional activity during this period, however. Emigration and the Colonial War made the greatest contributions towards integrating women in the labour market. Male-dominated emigration had a particularly important impact on the rural areas – the women who stayed behind in Portugal assumed the whole responsibility for managing both the family and the farm – and, despite the hard conditions in which they lived for many years, the situation helped to widen their independence quite considerably. The colonial war (Angola, Mozambique and Guinea-Bissau, 1961–74) caused the broadest repercussions, affecting all social strata. Mobilization into the armed forces gave rise to a serious deficit of male workers in all segments of the labour market. This in turn led to the massive recruitment of women and fostered their professional advancement appreciably.

Towards the end of the 1960s improved public social services demanded by growing urbanization also gave rise to quite a considerable increase in jobs for women. At the same time, as the large Portuguese entrepreneurial groups were consolidating their activities, men were offered more job opportunities, particularly in services to do with financing. At the start of the 1970s, gender-based tertiary labour segmentation revealed a clear split between the public and the private sector – a phenomenon still witnessed today.

The revolution that erupted in April 1974 unleashed a process of deepseated changes in Portuguese society and caused a radical break with traditional values (Gaspar 1987; Sousa Santos 1993) with resultant important changes being made in the level of women's employment and in their daily lives. After the revolution, concern about social inequalities became a fundamental issue. The new sociocultural values promoted social fairness, including the emancipation of women and the diffusion of new models of behaviour and anti-discrimination. The leadership for these changes did not, however, come from the feminist movements, which have always been poorly represented in Portugal, but rather from left-wing political forces working against a backdrop of party-political social life. The new Constitution, which came into effect in 1976, helped to eradicate innumerable gender-based legal discriminations, which had prevailed throughout the New State era, and also introduced the legal framework for ensuring equality between men and women. The framework was later consolidated by the passage of specific laws and today the basic aspects of juridical equality have been assured.

With a social order that oscillated between a welfare state and a socialist state, a marked growth occurred in public administration at both central and local levels and was responsible for an impressive growth rate in female employment during the post-revolutionary period; in 1981, the rate was the highest in southern Europe (Rodrigues *et al.* 1985). The state's role not only meant a significant increase in jobs for women, mainly in education, health and social assistance, but also assured the assistance of outside structures that were to help with the tasks of raising a family. With such help it became easier for women to be integrated into the labour market. Another outcome of the revolutionary process that was vital for women's emancipation was the democratization of education. Today, in terms of educational levels, girls have a clear advantage over boys (Cortesão 1988).

By contrast, the trade union movement's progress, which had been harshly suppressed during Salazar's dictatorship, induced some positive and some negative impacts on women's labour. The positive aspects include the overall improvement of working conditions. The negative ones have become obvious in the trade unions' discriminatory attitudes, which have attached more importance to protecting the so-called 'central' segments of the labour market dominated by men, and which have forgotten about the peripheral segments – thus permitting increasing instability in female employment. Male and female unemployment rates after 1974 (Figure 7.2) clearly show the trade unions' roles: in the serious economic depression during the second half of the 1970s, overall male unemployment dropped considerably despite the slump at home and abroad, which seriously hit certain segments of primarily male employment, such as metal works, the chemical industry and banking and financial services. At the same time, unemployment among women took a steep upward swing, even in sectors that were not so badly hit by the depression. This occurrence, however, failed to worry the trade unions to any great extent.

Throughout the 1980s, the rate of working women continued to climb fairly steadily and today it is among the highest in the European Union. Nevertheless, the rise has been accompanied by an increased instability, which affects all workers, although mainly the women. These conditions reflect liberalizing political guidelines and Portugal's joining the European Community (1986), both of which helped to promote the defensive, flexibility strategies that caused much job instability. Temporary jobs and contract work increased quite considerably, particularly in terms of women's labour. Because there is little social assistance and the standards of living are generally low, these strategies have meant great hardship for women; their working conditions have deteriorated and raising a family is fraught with even more difficulties. Of equal importance, since the mid-1980s, has been the process of rebuilding monopoly capitalism, partly as most of the firms which had belonged to the state's entrepreneurial sector during the revolutionary period were privatized. This phenomenon has caused important changes to be made in some sectors of employment. Changes have become particularly

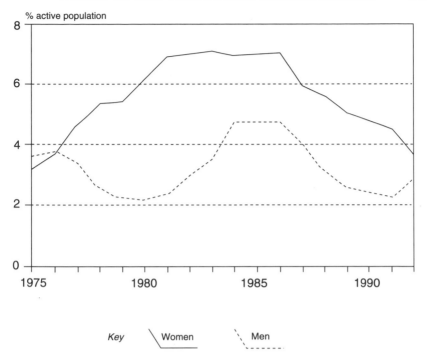

Figure 7.2 Unemployment rate, Portugal, 1975–92 (persons seeking new jobs)
Data source Instituto Nacional da Estadistica (INE), *Labour Enquiry*, 1975–92

apparent in financial services where efforts to increase productivity have generally taken precedence over strategies to cut production costs, with prejudicial effects on the female labour force. Management has adopted policies that mainly have to do with workers' availability and mobility – demands that accentuate gender differences.

THE DISTINCTIVE FEATURES OF WOMEN'S EMPLOYMENT IN PORTUGAL

Having summarized the basic factors leading to the contemporary configuration of women's employment in Portugal, I will now discuss some of the features that separate Portugal today from the rest of Europe, and even from other southern European countries. The distinctive nature of women's employment in Portugal has mainly emerged in four areas:

– in the numbers of working women;
  in the frequency of part-time jobs;
– in the composition of employment according to economic activity;
– in school education.

Numbers of women in the labour market

In 1992, women represented 44.3 per cent of the total Portuguese labour force and had an activity rate of 55 per cent (15 to 64 years of age), which is considerably higher than numbers registered in other southern European countries and situates Portugal among the top three members in the EU. In general, Portuguese women's employment has registered a deep, fast quantitative change over the last thirty years, though in order to define the process whereby they joined the labour market, it is necessary to examine the evolution of trends describing their activity (Figure 7.3). The impressive growth in women's employment noted in the 1960s resulted above all from the recruitment into the job market of young women (15 to 24 years of age). Up to the beginning of the 1970s, marriage and the birth of the first child placed major hurdles in the path of continued outside work, at least in the official market. It was a common practice among the less-privileged social groups, however, for married women to start doing paid domestic work at home or outside (as dressmakers, domestic helps, nursemaids and so on). The extent to which these activities were carried out cannot be assessed because their margin-alized economic position was overlooked by the official statistics. It was

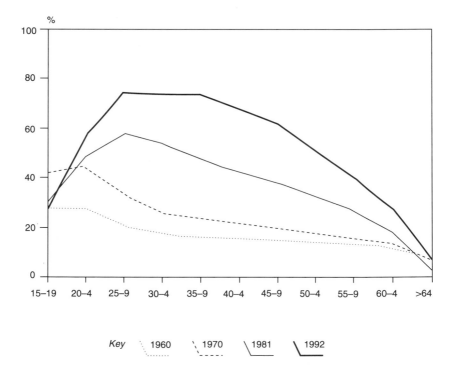

Figure 7.3 Activity rates of women by age groups, Portugal, 1960–92
Data source Instituto Nacional da Estadistica (INE), General Population Census, 1960, 1970 and 1981; *Labour Enquiry*, 1992

Table 7.1 Sectoral distribution of part-time women workers, Portugal, 1992

Agriculture	20.3
Industry	7.0
Commerce	9.3
Hotel industry	10.3
Transport/communications	2.0
Banks/finance services	7.3
Public administration	2.3
Social services	3.5
Domestic-personnel services	35.0

Source Instituto Nacional da Estadistica (INE), *Labour Enquiry,* 1992

mostly these young women who replaced men in the labour market due to the latter's absence in the Colonial War. As already mentioned above, the war and emigration were responsible for giving the most decisive impetus to women's employment over the past few decades and allowed women the chance of entering diverse segments of the labour market.

In the 1970s and 1980s, what could be called the settling-in period of women's labour occurred – marriage and motherhood ceased being major barriers against continuing employment. The pattern of female activity thus drew closer to that of men. Furthermore, women started to have a significantly longer active working life. Nevertheless, this evolution did not mean there was a major improvement in the organization of family chores. Nor did it suggest that progress witnessed in the state educational system or in social assistance (as Portugal moved towards becoming a welfare state) was on a sufficiently wide scale to explain the steep rise in women's labour. In other words, the majority of working women were still not able to share the burden of raising a family with either their husbands or the state. What happened was that these women had a double working day, alleviated somewhat by seeking the help of their original, extended families.

The transformation of women's activities in the last two decades basically derives from the predominant parameters of job opportunities. Portugal's peripheral position within the European economic framework justifies the continuation of cheap labour as one of the essential conditions for attracting investment. In fact, the survival of a

Table 7.2 Percentage of the female labour force in part-time work according to age groups, Portugal, Great Britain and the Netherlands, 1987

Age group	Portugal	Great Britain	Netherlands
14–24	7.3	20.5	39.6
25–49	8.9	50.2	62.8
50–64	14.7	54.9	70.0
Over 64	26.5	83.1	63.6

Source Eurostat, *Enquiry into the Labour Force,* 1989

large number of firms and the improvement of public services depend on not having to pay high wages or to demand extremely skilled workers. At a time when the trade union movement is exerting strong pressure for better working conditions, there is an increasing move to hire female labour, mostly situated out of reach of the unions. Indeed, for most entrepreneurs, this has become a fairly advantageous alternative. The considerable increase in women's education and their interest in the professional training programmes being developed and diversified are further factors enhancing the advantages of employing women.

At the same time, changing social and cultural values triggered off by the revolution have reinforced these directions. Woman as wife and as mother to her children are models being seriously questioned nowadays; the husband's interference in his wife's professional life has now been restricted by law and, more than anything else, by patterns of behaviour that are fast being accepted. The emancipation of women has made such a significant advance that it is manifested at sociocultural level.

Part-time work

Registered part-time work by women in Portugal is not very significant in numerical terms – only 11 per cent of women workers are thus engaged – and is mostly witnessed in agriculture and personal and domestic services (see Table 7.1).

The prevalence of part-time work across age groups is also distinctive in Portugal and differs considerably from patterns in those EU countries where it is more frequent (see Table 7.2). In Portugal, part-time work is mostly undertaken by the elderly population, while in the Netherlands or Great Britain it is a common practice among women over the age of 25 – clearly associated with motherhood. Values indicated in Table 7.2 show that in Portugal, part-time work does not have a decisive role in numerical flexibilization strategies affecting women's employment. This may be explained primarily because any ruling on the question would not benefit entrepreneurs who resort to hiring on this basis. Resorting to part-time jobs, however, has been associated with job instability; as a result, the chances of obtaining a fixed contract with this type of work are lower – 18.6 per cent of the women working part-time in 1992 did not have a permanent contract as compared with 13.7 per cent of women working full-time (Instituto Nacional da Estadistica [INE], *Labour Enquiry*, 1992). Overall, from the point of view of demand, part-time work is not very attractive in Portugal and the drop in wages probably dissuades women from seeking it. In most family units, the low standard of living goes against any decision to cut down on family income.

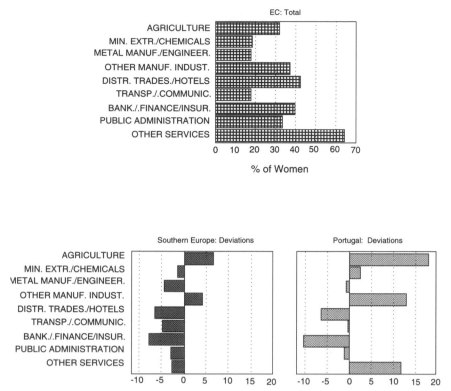

Figure 7.4 Women's participation in total employment by economic activities,
EC, Southern Europe and Portugal, 1987
Data source Eurostat, *Labour Force Survey*, 1987

The horizontal segmentation of labour

The presence of women in various economic activities in Portugal
displays a rather different pattern from that observed in the rest of the
southern European countries (Figure 7.4). Agriculture, the traditional
industries and domestic service still represent a large slice of women's
employment. Conversely, commerce and banking and financial services
register a lower female presence than in other countries in the European
Union. This horizontal segmentation of labour mainly reflects the
considerable importance of the traditional sectors within the overall
picture of the Portuguese economy. Resorting to female labour has
been the saving grace of a large number of firms which have managed
to stay afloat thanks to very low wages and the increasing instability of
contract work. The phenomenon is very common in agriculture,
although it is also witnessed in numerous industrial plants.

Table 7.3 School attendance rates according to age and gender, Portugal,
1960–92

Age group	10–14	15–19	20–4
1960			
Women	44.4	9.7	2.5
Men	51.1	12.7	4.2
1970			
Women	78.2	16.9	5.6
Men	82.1	18.1	5.4
1981			
Women	81.2	35.2	10.4
Men	84.0	30.4	7.8
1992			
Women	94.3	62.2	25.2
Men	94.9	52.0	17.9

Source Instituto Nacional da Estadistica (INE), *General Population Census*, 1970 and 1981;
INE, *Labour Enquiry*, 1992

School education

The level of women's education, particularly of girls', also illustrates another situation that is specifically Portuguese in nature. Contrary to what was observed above, however, Portugal has emerged in a more favourable light in this respect. As of 1974, overwhelming structural changes were implemented in the educational system. The school became a vantage point for consolidating democracy – the repressive ideology and traditional values that had left their marks on education during the New State were quickly eradicated; encouraging the most underprivileged classes to educate themselves became a priority target; all government schools became coeducational. During the 1970s, the school system seemed to have served political and ideological aims more than it heeded the need for economic development.

In the 1980s, the articulation between the school syllabus and occupational training once again emerged as an important issue – Portugal's membership of the European Community and the resulting changes effected in the home economy accentuated educational deficiencies. This state of affairs has prompted recent reforms within the lower- and upper-school systems and attempts have been made to re-establish teacher training. Furthermore, greater emphasis has been placed on technology and the exact sciences to the detriment of humanities and the social sciences.

Because of labour market pressure, and of simultaneous political, social and cultural changes, the level of women's education has risen quite considerably in the last couple of decades, and in recent years it has even inverted the previous status quo. Values presented in Table 7.3

suggest that there were two vital factors in the process of granting education to everyone. Mass school attendance was increased by phases: during the 1960s it covered basic primary school and early secondary school education; in the 1970s and 1980s it extended to full secondary school mass education; and in the last few years, a significant number of students have gone on to higher educational studies.

Making education available to the masses was accompanied by an intensive drive to promote women; today, the sex ratio stands in women's favour and may be explained by various factors, above all by job opportunities on the labour market that are more often than not directed at boys and young men; furthermore, making education available to the masses was accompanied by an intensive drive to reduce a family's giving preference to its sons' education over its daughters'. If trends in educational participation are compared with graduation rates of secondary school and university students (see Figure 7.5), the issue can be understood in greater depth. In former generations, males were more likely to complete secondary school education and to graduate from university than were females. Although education was somewhat easier to accomplish in the region of Lisbon than elsewhere, it was precisely in Lisbon that gender difference was at its peak. In recent years, however, the difference between males and females has clearly become inverted in the youngest age group. Once social and cultural restrictions inhibiting women from reaching the same educational levels as the men abated, education became women's main means of access to diverse segments of the labour market and, in particular, to the higher professional categories that are still placed beyond women's reach – although for other reasons. In a way, education has helped women to overcome some of the hurdles imposed on them by their household duties, thus allowing them to pursue their careers.

Nevertheless, it is worth noting that even though education became more democratic as a result of the 25 April 1974 Revolution, as is expressed so well in the profile of university education (see Figure 7.5) where the more numerous presence of women comes as a later repercussion, the highest university graduation rate is recorded for men in the 35–9-year-old age group and for women in the 30–4-year-old age group. The imposition of restrictions on access to university education at the start of the 1980s has also presented a relative advantage to women.

The intensive feminization of higher education has affected all scientific spheres, although the differences between the various study areas continue to reproduce basic features in the traditional patterns of male and female education (see Table 7.4). In 1970, there was a deep gender-based split in university education – women were concentrated in the arts and pharmacy faculties while the other colleges, particularly engineering, veterinary medicine and law, registered a large percentage of men. In the last two decades, this distribution has drastically changed; by 1990, the only field in which men had a clear majority was engineering.

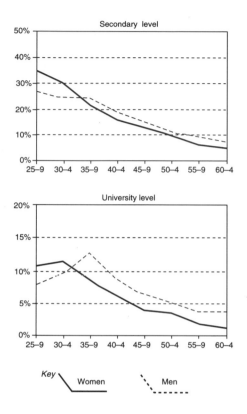

Figure 7.5 Education levels by age groups, Portugal, 1992
Note Secondary level includes upper levels.
Data source Instituto Nacional da Estadistica (INE), *Labour Enquiry*, 1992

It is important to note, however, that the increasingly more numerous presence of women in scientific areas traditionally reserved for men – law, medicine and veterinary medicine – did not prompt an inverse movement of men into arts courses and pharmacy. This evolution singles out an aspect that is common to other fields, namely, that the road to equality normally goes one-way – women 'invade' masculine strongholds but men refrain from occupying women's spaces to any marked extent. This tendency suggests that the dominant male models have always been present in the process of social equality, which means that the patriarchal relationship still persists.

Table 7.4 Percentages of women enrolled in colleges at the Universities of Coimbra, Lisbon and Oporto (government education)

	1970	*1980*	*1990*
Coimbra University			
Faculty of Science	36.9	39.3	43.0
Faculty of Law	20.1	23.9	51.8
Faculty of Arts	80.3	70.1	77.7
Faculty of Medicine	26.1	43.8	50.5
Faculty of Pharmacy	79.3	63.5	75.3
Faculty of Economics	–	32.6	46.4
Lisbon University			
Faculty of Science	41.5	63.3	59.1
Faculty of Law	19.5	32.8	53.9
Faculty of Arts	81.8	74.6	78.8
Faculty of Medicine	37.2	50.6	56.6
Faculty of Pharmacy	81.0	71.4	79.3
Faculty of Psychology/Educational Sciences	–	–	80.0
Lisbon Technical University			
Higher School of Veterinary Medicine	12.4	26.4	52.4
Higher Institute of Agronomy	27.7	40.2	56.0
Higher Institute of Economy and Management	25.7	33.2	45.2
Higher Institute of Social/Political Science	28.6	39.7	71.6
Higher Technical Institute	15.0	14.8	17.0
Faculty of Human Motoricity	–	27.3	47.5
Faculty of Architecture	–	–	36.6
New Lisbon University			
Faculty of Medicine	–	66.3	60.6
Faculty of Human/Social Sciences	–	41.7	74.7
Faculty of Sciences	–	52.9	43.9
Faculty of Economics	–	42.5	45.5
Oporto University			
Faculty of Science	39.7	69.2	62.3
Faculty of Economics	26.1	34.0	43.2
Faculty of Engineering	6.9	14.4	17.7
Faculty of Pharmacy	87.8	68.2	74.7
Faculty of Arts	73.9	75.0	68.0
Faculty of Medicine	32.5	47.2	60.6
Higher Institute of Biomedical Sciences	–	55.2	60.0

Source Instituto Nacional da Estadistica (INE), *Educational Statistics*, 1970, 1980 and 1990

PORTUGUESE WOMEN: AT THE PERIPHERY OR IN THE CENTRE?

Women's place on the fringe of the labour market is common to various models of development, although the models take on distinctive forms in different countries, depending upon the position they occupy in the international division of labour and on the specific social and cultural idiosyncrasies that emerge from particular historical processes. In Portugal, the migration process of the 1960s and the Colonial War in Africa (1961–74) made it possible for many women to find full-time jobs on the labour market. Thus, in some respects, Portuguese women could be described as 'central' in the labour market. Yet Portugal's semi-peripheral character has also shaped women's work and daily lives. The continued existence of a large traditional economic sector based on extensive accumulation even today gives rise to a greater demand for female labour than occurs in countries of the 'centre'. Simultaneously, the strong ties that have continued to link members of the extended family (a typical feature in the semi-peripheral countries) have frequently facilitated the provision of help with the family's domestic chores that has made life easier for the professional working woman in the last couple of decades.

Yet women's increasingly greater access to the sectors of employment that require relatively higher qualifications, especially where educational qualifications are an important criterion (doctors, lawyers, university teachers and so on), reflects not so much peripherality as the radical changes brought about by the 1974 revolution: on the one hand, cultural values and social forms were torn apart and forced to become less traditional and much more permissive; on the other hand, the democratic spread of education allowed increasingly more women to educate themselves at all levels, so that among young adults, women now have a significant advantage over the men in terms of educational qualifications, even while they continue to experience some forms of occupational segregation.

REFERENCES

André, I. (1991) 'The employment of women in Portugal', *Iberian Studies* 20, 1–2: 28–41.

—— (1994) 'O falso neutro em geografia humana. Género e relação patriarcal no emprego e no trabalho doméstico', unpublished PhD dissertation in Human Geography, Lisbon University.

Beleza, M. T. (1990) 'Mulheres, direito, crime ou a perplexidade de Cassandra', unpublished PhD dissertation, Faculty of Law, Lisbon University.

Cortesão, L. (1988) *Escola, Sociedade: Que Relação?* Oporto: Afrontamento.

Ferrão, J. (1985) 'Recomposicao social e estruturas regionais de classes', *Análise Sociale* 21, 87–9: 565–604.

Gaspar, J. (1987) *Portugal: Os Próximos 20 Anos,* Lisbon: Fundação Calouste Gulbenkian.

Lipietz, A. and Leborgne, D. (1992) 'Flexibilité offensive, flexibilité défensive: deux stratégies sociales dans la production de nouveaux espaces économiques' in G. Benko and A. Lipietz (eds) *Les Regions qui gagnent. Districts et réseaux: les nouveaux paradigmes de la geographie économique*, Paris: PUF.

Rodrigues, E. F., Figueredo, C., Cordoril, F., Ribeiro, J. F. and Fernandes, L. (1985) 'Especialização internacional, regulação económica e regulação social, Portugal: 1973–1983', *Análise Sociale* 21, 87–9: 437–71.

Rodrigues, M. J. (1988) *O Sistema de Emprego em Portugal: Crise e Mutações*, Lisbon: Publ. D. Quixote.

Silva, M. (1984) *Retorno, Emigração e Desenvolvimento Regional em Portugal*, Lisbon: IED 8.

—— (1993) *O Emprego das Mulheres em Portugal: A 'Mão Invísivel' na Discriminação Sexual no Emprego*, Oporto: Afrontamento.

Sousa Santos, B. (1993) 'O Estado, as relações salariais e o bem-estar social na semi-periferia: o caso português', in B. Sousa Santos (ed.) *Portugal: um Retrato Singular*, Oporto: Afrontamento.

Walby, S. (1986) *Patriarchy at Work*, London: Polity Press.

—— (1990) *Theorizing Patriarchy*, Oxford: Blackwell.

8

CONTRASTING DEVELOPMENTS IN FEMALE LABOUR FORCE PARTICIPATION IN EAST AND WEST GERMANY SINCE 1945

Jürgen Schmude

The reunification of Germany in 1990 has highlighted the implications of the nature of the economic system and of state policies for the extent and nature of women's participation in the workforce. It has also demonstrated the persistence of gender-based occupational segregation. In this chapter I will examine the evolution of women's employment in the Federal Republic of Germany and the German Democratic Republic since the Second World War, focusing on the changing rates of participation, the distribution of women's jobs across sectors and the regional variations within each of the territories. I will conclude with an assessment of the prospects for women's employment in the 'new' Germany.

PRELIMINARY REMARKS AND HISTORICAL FOUNDATIONS

The radical economic transformation resulting from the industrial revolution of the nineteenth century brought in its wake far-reaching changes in the lives of women of all social strata. Women who had previously been accustomed to performing their work at home, had, now at least in part, to work in factories, since the productivity of machines was greater than that of manual outwork. Women of different classes responded to the new conditions in different ways. Proletarian women employed in the factories formed working women's associations with the goal of improving their status. By contrast, women of the bourgeois middle class, who had previously also been active for the most part within the household and had made a considerable share of their own household goods (such as foodstuffs and clothing) were increasingly deprived of their work by new and cheaper forms of mechanically based production. This inevitably led many of them to demand opportunities to work outside the home (Beilner 1971: 11). In particular, single women who were not interested in entering into marriages of economic convenience needed to find gainful employment (see, for example, Twellmann 1972: 29). Such opportunities were provided by the opening to women of occupational fields that had been previously inaccessible to them or to which their access had been strictly regulated. The teaching profession serves as a typical example in this context.

Women were admitted into the teaching profession at *Volkschulen* (later termed *Grund-* and *Hauptschulen*; that is, roughly, elementary and junior high schools) in the second half of the nineteenth century in individual states of what was later to become the German Empire. This access was initially regulated by various legislative measures ranging from a proviso making a minimum school size (of at least three teachers) a precondition for the employment of female teachers to a restriction requiring women teachers to be single (celibate). The immediate result of these interventions was to create a disparity between the centre and the periphery in the percentage of female teachers on school staffs (see also Schmude 1988). Thereafter, in spite of a gradual change in basic societal norms (which produced an increased acceptance of female workforce participation), these basic legislative regulations were loosened or tightened according to the current situation on the labour market. For example, celibacy was abandoned in times of teacher scarcity (such as during the First and Second World Wars), but reinstated during crises of over-capacity (such as during the economic crisis in the Weimar Republic). In this way, women were put into the role of reserve force in the labour market.

All in all, one can conclude that female workforce participation up until the end of the Second World War was concentrated in a few occupational fields and its overall volume was more or less strongly influenced by state planning measures which varied in strength according to the given economic situation. After the Second World War, the division of Germany led to very different defining conditions on the East and West German labour markets. Accordingly, the courses of development in female workforce participation differed markedly between the two German states. Differences also exist in the statistics drawn up on workforce participation in the two states and, consequently, in the available databases, in terms both of individual definitions and of the system and times of data collection. Moreover, not all of the data collected in the German Democratic Republic (GDR) were published. Consequently, direct inner-German comparisons are methodologically questionable. For this reason, the developments of female workforce participation for the period from the Second World War until reunification in 1990 will be treated separately for each of the two areas of Germany.

THE POSTWAR DEVELOPMENT OF FEMALE WORKFORCE PARTICIPATION IN THE GERMAN FEDERAL REPUBLIC

General development

As in other industrialized European nations, the workforce participation of women increased in the German Federal Republic (FRG) after the Second World War. Shortly after the war, about 7.2 million women were active (accounting for 35.6 per cent of the entire workforce). By 1987, they made up 38.1 per cent of this workforce (see Table 8.1).

The rate of employment among women (defined as the number of working women as a percentage of the total female population between 15 and 65 years of age) rose even more steeply. Whereas in the years following the Second World War this rate reached only 40 per cent, it rose considerably thereafter, especially starting in the 1960s, ultimately reaching 55.5 per cent by 1989. However, this rise is not based on a continuous development; instead, there were several differently motivated phases of expansion in female workforce participation.

The first postwar decades were marked by reconstruction and a quickly emerging restructuring of the industrial sector. The available male labour force was insufficiently large and, even with the inclusion of women, the labour demands of the 1960s could not be met. Instead, there was an extensive recruitment of foreign labour (especially in the secondary sector), particularly since the greatest period of labour demand coincided with the highpoint of the West German baby boom. It was at this time that the phenomenon of part-time work first increased in quantitative importance, since the familial situation of married women partly limited their labour market availability.

In 1961, clear regional disparities existed in the female percentage of the total workforce as illustrated by a comparison of the different *Länder* (hereafter referred to as states) making up the Federal Republic. The difference between states was approximately 15 per cent, considerably greater than that found in the GDR (compare Figure 8.1a with Figure 8.5a). These regional differences were primarily caused by the different

Table 8.1 Total workforce and women's share of this workforce according to economic sector, Federal Republic of Germany

Sector	1957		1970		1987	
	No. ('000s)	*Female %*	*No. ('000s)*	*Female %*	*No. ('000s)*	*Female %*
Primary	4,112	54.6	2,370	52.7	866	34.8
Secondary	12,079	25.3	12,797	24.9	11,247	23.8
Tertiary	9,462	44.6	11,287	44.9	14,794	49.2
Total	25,653	37.1	26,454	36.0	26,907	38.1

Sources Statistisches Bundesamt 1971 126 ff.; 1989: 94 ff.; table compiled by author

sectoral structures of the economies of the individual states. The states of Saarland and North Rhine-Westphalia were marked by the large relative sizes of the secondary sectors in their economies (see Figure 8.1a). The branches that are now termed old industries then played an important role (especially coal and steel). Since precisely these branches displayed the lowest rates of employment among women, the lowest percentages in overall female workforce participation resulted here (29.3 per cent in Saarland and 32.3 per cent in North Rhine-Westphalia). In contrast, the markedly higher levels of female workforce participation in the states of Bavaria and Baden-Württemberg can, on the one hand, be accounted for by the considerably greater importance of the primary sector in total employment there. On the other hand, the higher rate also reflected the contrasting composition of the secondary sector in these states (for instance, the greater importance of the textile industry in Baden-Württemberg). At the beginning of the 1960s, the service sector was still smaller than the secondary sector in terms of number employed (12.7 million in the secondary sector versus 10.2 million in the tertiary sector). Moreover, the percentage of the total tertiary-sector workforce made up

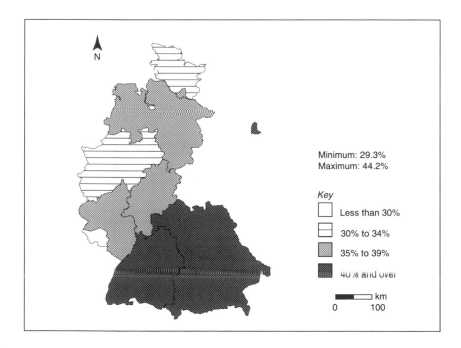

Figure 8.1 a) Women as percentage of the total workforce, Federal Republic of Germany, 1961
Source Statistisches Bundesamt 1964 30ff; figure by author

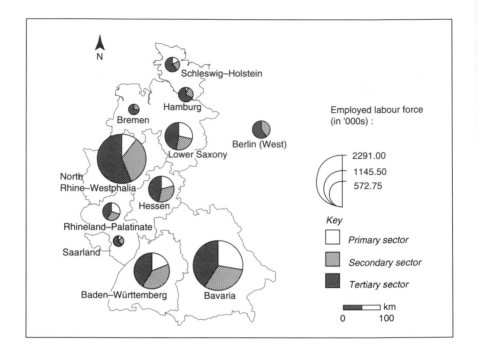

Figure 8.1 b) Breakdown of the workforce according to economic sectors,
Federal Republic of Germany, 1961
Source Statistisches Bundesamt 1964 30ff; figure by author

of women varied only marginally between individual West German
states; accordingly, this was not a source for the regional differences in
female workforce participation.

In addition to changes in society, shifts in the relative importance of
economic sectors in the West German economy were also responsible for
the further increase in the rate of employment among women in the
1970s and 1980s. After the reconstruction phase of the 1950s and 1960s
with its heavy emphasis on the secondary sector of the economy,
industrial employment started to stagnate in the 1970s, and employment ultimately dropped in the 1980s. At the same time, the service
sector started to expand rapidly. Whereas the secondary sector still
accounted for almost half of the entire workforce at the end of the
1950s, in the course of the 1970s and 1980s it was overtaken by the
tertiary sector in absolute numbers (see Table 8.1). This sectoral restructuring of the economy did not affect men to the degree that it affected
women, for even in 1987 there were still more men employed in the
secondary sector than in the service sector. In contrast, the service sector,

which is very heterogeneous in terms of the qualifications and the labour-time demands it makes, offered women an array of new employment opportunities. A disproportionate growth in the importance of this sector resulted, especially in terms of the absolute number of jobs provided for women. Correspondingly, it is only in this sector that the relative share of women in the total workforce increased between 1957 and 1987 (see Table 8.1).

The 'tertiarization' of the structure of the economy led to a levelling of regional differences between different German states in female workforce participation (compare Figure 8.2a with Figure 8.1a), for there was much less variation in the relative share of women employed in the service sector than in the secondary sector. By 1987, a difference of less than 10 per cent existed between individual German states in the minimum and maximum relative share of women as a percentage of the total workforce (1961: 15 per cent).

Finally, the years from 1985 until 1990 were marked by a growth in the size of the workforce of about 1.4 million, almost two thirds of which were women. In 1990, the rate of employment among women had risen

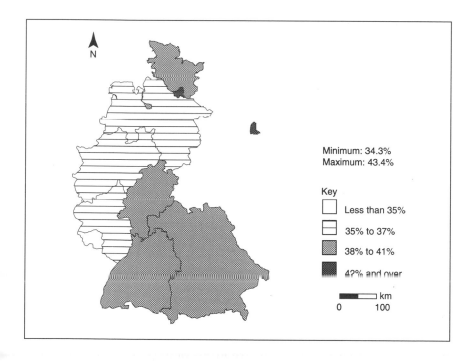

Figure 8.2 a) Women as percentage of the total workforce, Federal Republic of Germany, 1987
Source Statistisches Bundesamt 1990 40ff; figure by author

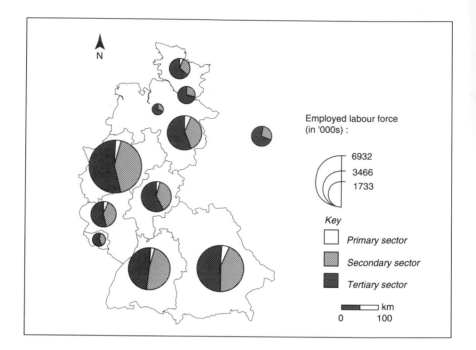

Figure 8.2 b) Breakdown of the workforce according to economic sector,
Federal Republic of Germany, 1987
Source Statistisches Bundesamt 1990 40ff; figure by author

to around 55 per cent (from 47.2 per cent in 1961). However, a
considerable proportion of these positions were part-time in character.
Nevertheless, women were held to be the 'winners on the labour market'
in this period (Engelbrech 1991: 20). In spite of this positive development, the employment situation of women at the end of the 1980s
continued to be beset by problems. For instance, they continued to be
disproportionately hard hit by unemployment.

Sociodemographic aspects and extent of workforce participation and unemployment of women in the FRG

The increase in workforce participation after the Second World War did
not occur uniformly across all subgroups of women; instead, there were
essential differences according to family status and age cohorts. Above
all, married women were available in greater numbers at the time of
increased labour demand at the end of the 1950s. Accordingly, it is the
rates of unemployment among married women that have risen most

sharply since the end of the 1950s, and the interruption of one's work history, which had previously occurred at the time of marriage, was now postponed to a later date, if it took place at all. Not only did women less frequently interrupt their work histories, but they also markedly curtailed the length of any hiatus. Moreover, it became increasingly common for women in the older age groups, who had often interrupted their work-force participation for a family-oriented phase (birth and/or raising of children), to attempt to return to occupational life. Thus, for example, the rate of employment among women between the ages of 35 and 45 increased from 38 per cent to around 50 per cent from 1977 to 1987. The classic three-phase hypothesis of Myrdal and Klein (1956), which describes a clearly recognizable decrease of female workforce participation during the period of childcare, became increasingly less valid as a description of typical female work histories, not only in other West European countries (and especially in Denmark and France) but also in Germany, and this is a continuing trend. Instead, women in the FRG entered the family-oriented phase at increasingly older ages, and became either less likely to interrupt their employment, or apt to take markedly shorter breaks. A clearly reduced average number of children per woman and the expansion of part-time work were the basic preconditions for this development.

The FRG belongs to those countries in Europe (as do Greece, Spain, Ireland, the Netherlands, and Great Britain) in which the degree of female workforce participation is decisively affected by whether the woman in question has a child or not. However, it is increasingly found in the FRG that the first child 'only' leads to a reduction in the extent of workforce participation (from full-time to part-time), whereas in Great Britain the classic three-phase theorem continues to be valid (see, for example, Helberger 1988: 737). Just the 'existence' of one child in a West German family in the 1980s, however, continued to have a clear effect on female workforce participation or on the extent of this participation. By contrast, in Italy and Portugal this effect initially takes place with the second child, and in France it does not even occur until a third child is born (see Knauth 1992: 16). An important cause of this difference lies in opportunities for non-family childcare (day-care centres for infants and pre-school children) that are markedly fewer in the FRG than, for instance, in France (écoles maternelles starting at age 2). The basic conditions of care for children in case of illness were also much more restrictive in the FRG than in the GDR (see Shaffer 1981: 19). As expected, there is still a tendency for the workforce participation of women to fall as the number of their under-age children rises. And the large majority of working mothers are employed part-time.

The expanded opportunities for part-time work are the result of the increased importance of the service sector, for, according to Engelbrech (1987), part-time employment was provided above all in small, non-industrial-sector companies with fewer than 100 employees. Even today, this part of the economy accounts for about three quarters of all part-time positions, with the latter concentrated in commerce and in the hotel and restaurant business, as well as in unskilled service occupations. Just from March 1988 until March 1993, the number of part-time positions increased from around 2.1 million to 2.8 million. In March 1993, part-time positions accounted for 11.9 per cent of all positions, with a ratio of thirteen women for every man in such employment (see Bundesanstalt für Arbeit 1993). Due to the relatively large proportion of part-time positions, the percentage of paid labour hours accounted for by women increased only from about 33 per cent in 1960 to 36 per cent in 1988. Nevertheless, in terms of demand, part-time positions for women remain scarce today. What has to be realized is that women in their family-oriented phase who sought part-time positions were especially hard hit by unemployment. Thus, for example, in 1988, the ratio between the supply of part-time positions and the demand for them on the part of unemployed women was about 1:10 (Amtliche Nachrichten der Bundesanstalt für Arbeit [ANBA] 1989). At that time, about 16 per cent of all women in the FRG with two children were employed full-time and 24 per cent part-time (Deutsches Institut für Wirtschaftsforschung [DIW] 1990).

In spite of the greater opportunities of access to the labour market afforded women, they are increasingly affected by unemployment. In the time of full employment in the 1960s, the unemployment rate of women was always lower than that of men (both under 1 per cent), but with the generally rising rates of unemployment since the beginning of the 1970s, women have been more greatly affected than men. By the decade's end, the female unemployment rate of approximately 5 per cent was about 2 per cent higher than that of their male counterparts. This difference remained more or less constant until the end of the 1980s, although on a higher overall level. In 1989, the 8.9 per cent unemployment rate of women was again about 2 per cent higher than that of men. Moreover, women's average duration of unemployment was longer than men's and their rate of re-employment was also markedly lower than the comparable figure for men (ANBA 1991). Women with low qualifications were especially hard hit by unemployment.

Sex-specific qualifications and inner-occupational status in the FRG

The postwar increase in the rate of employment among women varied according to educational attainment. Whereas the traditionally low rates of employment among women with low levels of qualification did not rise, the higher the qualification, the greater the percentage of women working. A fundamental precondition here was the gradual convergence of the educational levels of men and women after the Second World War. Thus, by the middle of the 1970s, about 50 per cent of those who had completed the *Abitur* (school-leaving examination and university entrance qualification) were women, whereas shortly after the Second World War, women accounted for only about 30 per cent of this group. A similar development can also be found in the university sector, though here the proportion of women still varies sharply according to discipline. Women are clearly overrepresented in the humanities, but they remain a minority in the natural sciences.

A clearly sex-specific pattern can be recognized in the composition of different occupations or occupational groups, even as the relative share of women has continuously increased (as in commerce). The concentration of female workforce participation in the service sector has resulted in the 'feminization' of individual occupations. The strong concentration of women's occupational activities – and thus also of their occupationally specific training – in a few occupational fields is vividly demonstrated by the fact that 78 per cent of all gainfully employed women were concentrated in fifteen occupations or occupational groups in 1989 (in contrast, the fifteen most common male occupations accounted for 'only' 57 per cent of all gainfully employed men).

Furthermore, within the different occupational groups there is a vertical differentiation in the relative shares of each sex. Analyses of status within occupations provide evidence of this. Women 'work much

Table 8.2 The fifteen most common female occupations or occupational groups in 1989

Occupation or occupational group	Women employed in occupation %	Female workforce in occupation %
Skilled or assistant office workers	68	24
Merchandise-sales personnel	62	12
Health-care occupations other than below	85	8
Cleaning professions	84	4
Occupations in commercial bookkeeping and data processing	53	4
Nursing, therapeutic and care professions	80	4
Teachers	48	3
Untrained workers without degrees	36	3
Agricultural workers	79	3
Trained banking and insurance personnel	43	3
Hotel and restaurant personnel	62	2
Textile-processing workers	89	2
Personal-hygiene-related occupations (hair stylists, manicurists, cosmeticians, etc.)	86	2
Cooking-related occupations	63	2
Entrepreneurs, organizational experts, auditors	21	2
Total		78

Source After Maier 1993: 267

more frequently in more inferior, more burdensome, and less well-paid positions, which are also connected to lesser opportunities for promotion within one's occupation' (Maier 1993: 268). Accordingly, even at the end of the 1980s, considerable differences still existed in the average incomes attained by men and women, though these differences were markedly smaller than those of the years immediately following the war. Between 1960 and 1980, the average income of women rose from 65 per cent to 72 per cent of the income of men (Helberger 1988: 740). There were, however, industry- and qualification-specific differences here. Thus, in the West German textile industry, women attained 81.1 per cent of average male wages in 1978, but in the food, beverage and tobacco industry only 69.9 per cent. On the whole, the sex-differentiated distributions of net income demonstrated a median that was for women clearly lower (for example, in 1978: DM 800–1,000) than for men (1978: DM 1,400–1,800). The sex-specific differences in income are still recognizable today and industry-specific differences continue to exist (see Maier 1993: 272). The income gained from employment by women was, however, not as important in the FRG as in the GDR, since men's income largely sufficed to cover family living expenses. Moreover, the tax- and family-policy elements in the system of retirement benefits are geared towards the one 'breadwinner' model of the family (see DIW 1990: 577).

The higher level of education of women after the Second World War did not eliminate vertical segregation, but simply shifted it on to a higher level of qualification. And here, in postwar Germany, the relative proportion of women in individual occupations changed visibly not only over the course of time, but also in terms of its regional distribution. Teaching provides a good example of this situation; today it is frequently termed a typically 'woman's occupation'.

The feminization of an occupation: the woman teacher

Since teaching – at least in the field of general education – can be considered a ubiquitous occupation, it is especially well suited to the spatial-temporal analysis of the advance of women into a field previously dominated by men (see, for example, Schmude 1988; or Meusburger and Schmude 1991). For reasons of research economy, the following presentation limits itself to an analysis of events in the West German state of Baden-Württemberg. However, processes similar to those depicted here took place in the other West German states.

Women made up about 33 per cent of the teaching staff at *Volkschulen* (roughly, primary schools) in Baden-Württemberg after the Second World War. Within only thirty years, this figure reached well over 50 per cent (1975: 57 per cent), ultimately leading to general talk of a 'feminization of the teaching profession'. Indeed, this was the first time since the introduction of mandatory education that male teachers were actually replaced by women teachers.

On the one hand, the rapid increase in the proportion of women on the teaching staff was related to the general increase in female workforce participation after the Second World War. On the other hand, the profession of primary school teacher suffered a significant loss in prestige that reduced its attractiveness. Since no university education was necessary for teaching at primary schools until the 1950s (with teachers being trained at teaching colleges [*Lehrerseminare*] instead), the primary school profession served as a 'launching pad' for the intergenerational process of upward mobility: it was the first rung on the mobility ladder into the academic world (see, for example, Recum 1955; or Bungardt 1959). When the requirement for primary school teaching was changed to studies at a *Pädagogische Hochschule* (making primary school teaching preparation more a part of the university educational system), entry into the profession became almost as difficult as preparation to be a *Gymnasiallehrer* (roughly, a teacher at a high school preparing for university studies), even though the latter position was ranked much higher on the occupational prestige scale. Thereafter, men increasingly left primary teaching to women and concentrated on the occupational field of *Gymnasiallehrer*, which they continued to dominate. The latter was never feminized. At its maximum, the percentage of women on *Gymnasium* teaching staffs in Baden-Württemberg never exceeded 30 per cent (1975).

The feminization process in the primary schools (*Volksschulen*, equivalent to their later replacements, *Grund-* and *Hauptschulen*, or, roughly speaking, elementary and junior high schools) demonstrated strong centre–periphery disparities which existed until the 1970s. Differences in women's relative representation in town and country resulted from 'state decrees' that basically made a minimal school size a precondition for the acceptance of women teachers. Even after these regulations were dropped, however, differences continued to exist. The regression-analytical model (Figure 8.3) of the relationship between community size and the proportion of women on *Volksschule* teaching staffs (and later on *Grund-* and *Hauptschulen* teaching staffs) shows a practically linear relationship for 1911, 1951 and 1975 (r^2 being over 0.9 for each). The disparities between centre and periphery even increased between 1911 and 1951 and remained constant, on a higher level, between 1951 and 1975. A primary cause of these disparities was the persistent lack of acceptance of women as teachers in rural areas. In a considerable number of rural schools, lessons were held in one classroom for several age groups, a situation commonly felt to place excessive demands on women teachers. Women were also rejected because teachers in rural communities often had to perform other important duties that were typically considered part of the 'men's domain' (such as that of community secretary, head of the fire brigade, and/or organist).

Only after 1975 did this strict discrepancy between the centre and

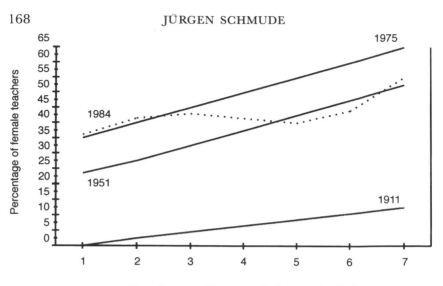

Figure 8.3 The relationship between the percentage of women on the teaching staffs of *Grund-* and *Hauptschulen* (full-time and full-year teachers) and the size of the community in which the school is located
Source Schmude 1988: 87

periphery in the percentage of female teachers begin to disappear, partly because many small, peripherally located schools were closed in favour of larger, more central, regional schools (*Schwerpunktschulen*). Finally, by 1984, a linear relationship could no longer be established between the percentage of women on the teaching staffs of *Grund-* and *Hauptschulen* and the size of the community in which the school was located. Nevertheless, in the same period of time that these centre–periphery disparities were being eliminated, the percentage of women teachers dropped once again as a result of an overcapacity crisis in the teaching profession triggered by the generally problematic employment situation.

The overcapacity crisis in the teaching profession from the end of the 1970s and especially in the 1980s, which also applied to teaching positions at secondary schools, made teaching at primary and junior high schools once again more attractive for men. The relative share of male teachers rose and, by 1990, the percentage of full-term female teachers had dropped to 40.2 per cent. This was caused by the greater attention paid to male candidates in teacher recruitment and the fact that women were frequently overproportionately represented among teachers leaving the staff (age-based retirement, family-related interruptions of service, reductions to part-time teaching).

The representation of women teachers has also been affected since the mid-1970s by a marked expansion of part-time teaching. By 1990, 42 per cent of all teachers working at *Grund-* and *Hauptschulen* were employed part-time, with 90 per cent of these positions occupied by women. This is one of the major reasons why the teaching profession is considered an ideal occupation for mothers.

The conditions at *Grund-* and *Hauptschulen* sketched here, however, do not apply to other teaching positions at other institutions, either secondary schools (especially *Gymnasien*) or universities. The higher the level of the educational institution, the lower the percentage of women on the teaching staff. This becomes especially clear in the university domain. Women accounted for 17.1 per cent of all lecturers and assistants in 1990 and only 5.5 per cent of all professors. Analogously, the higher the educational degree, the lower the percentage of women: in 1990, women accounted for 36.5 per cent of all university examinations completed, 27.7 per cent of all doctorates and 9.9 per cent of all 'habilitations' (see Figure 8.4).

THE POSTWAR DEVELOPMENT OF FEMALE WORKFORCE PARTICIPATION IN THE GERMAN DEMOCRATIC REPUBLIC

General development

Female workforce participation was greater in the GDR than in the FRG after the Second World War. By 1956, women already made up 43.6 per cent of the total workforce and in the following years their overall share continued to increase (Table 8.3). The secondary sector had a lower proportion of women than the tertiary sector. Unlike its western neighbours, in the GDR the secondary sector continued to increase in importance even after 1961, such that in 1986 a little over 50 per cent of its entire workforce was employed in this area. Thus, one cannot speak of a tertiarization of the GDR economy through the end of the 1980s. This situation in turn limited the rise of female workforce participation, which would have been even greater because of the high percentage of

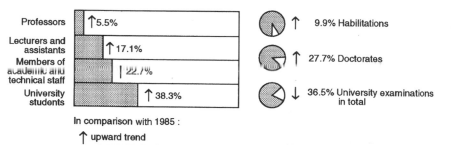

Figure 8.4 Percentage of women in different teaching professions and at different levels of educational achievement
Source Statistisches Bundesamt 1987 and 1991; figure by author

Table 8.3 Total workforce and women's share of this workforce according to economic sector, Federal Republic of Germany

Sector	1956		1970		1987	
	No. (*000s)	Female %	No. (*000s)	Female %	No. (*000s)	Female %
Primary	1,684	49.0	997	45.8	927	38.5
Secondary	3,779	33.6	3,995	38.7	4,311	38.7
Tertiary	2,715	54.1	2,777	63.0	3,310	65.7
Total	8,178	43.6	7,769	48.3	8,548	49.1

Source SVZ 1958: 190 ff.; 1971 60 ff.; 1988: 120 ff.; table compiled by author

this sector's workforce made up of women (1986: 65.7 per cent). The primary sector had lost importance both in absolute terms of overall employment and in relative terms of the workforce participation of women.

The growing integration of women into the labour market is also evidenced by the clear rise in sex-specific rates of employment. In

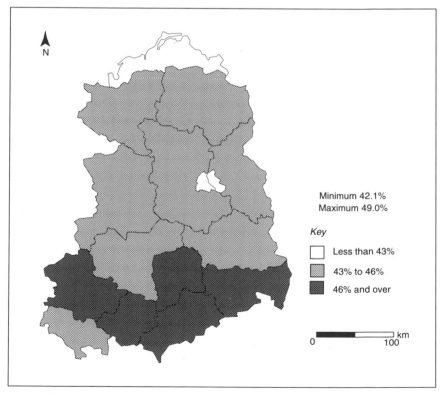

Figure 8.5 a) Women as percentage of the total workforce, German Democratic Republic, 1961
Source Statistisches Jahrbuch der DDR 1962: 179ff; figure by author

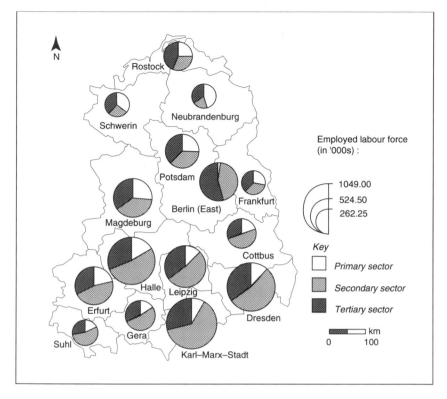

Figure 8.5 b) Breakdown of the workforce according to economic sector,
German Democratic Republic, 1961
Source Statistisches Jahrbuch der DDR 1962: 179ff; figure by author

1969, the rate of employment among women between the ages of 25
and 60 had already reached 73.4 per cent (as compared to 45.1 per cent
in the FRG), and by 1984, it was 84.5 per cent (56.1 per cent in the
FRG). It stayed at this high level until reunification (DIW 1990: 576).
These high rates of employment among women were also only marginally affected by the temporary release of women workers for maternity
leave. Thus, the high rate of employment among women in the GDR
was similar to that of the Soviet Union, where, as early as the beginning
of the 1960s, there were no longer any sex-specific differences in rates of
employment (Helberger 1988: 738). In the European Community,
comparably high rates of employment were reached only by men. For
example, even in 1990, 79 per cent of all women between the ages of 25
and 29 were employed in the GDR.

The regional disparities in the workforce participation of women were
on the whole also much smaller than in the FRG. By the beginning of the
1960s, women already made up 42 per cent or more of the total
workforce in almost all administrative districts (Figure 8.5a) and had

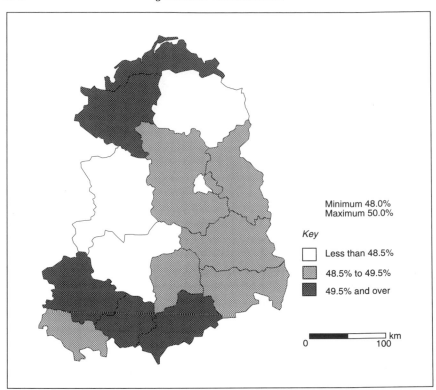

Figure 8.6 a) Women as percentage of the total workforce, German Democratic
Republic, 1987
Source Statistisches Jahrbuch der DDR 1987: 120ff; figure by author

surpassed the highest percentages reached in the West German states
(Figure 8.1a). Participation rates as high as 49 per cent were reached in
those districts today found in the states of Saxony, Thuringia and the
western part of Brandenburg. The primary cause of this distributional
pattern was the concentration of branches of industry that employ a high
proportion of women in their workforces (such as the textile industry) in
Saxony and Thuringia (above all in the districts of Erfurt, Gera, Karl-
Marx-Stadt, Leipzig and Dresden). The districts of West Brandenburg
reflected more than anything else the above-average proportion of
women in the service sector.

These relatively small disparities were reduced even further by the
increasing participation of women in the workforce. In 1987, a cross-
district comparison exhibited only a little less than a 2 per cent difference
between the largest and smallest percentages (Figure 8.6a). Apparently,
by this time, differences in economic structure no longer had any major
influence on female workforce participation.

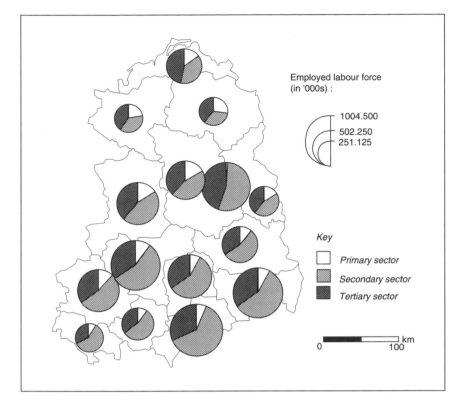

Figure 8.6 b) Breakdown of the workforce according to economic sector,
German Democratic Republic, 1987
Source Statistisches Jahrbuch der DDR 1987: 120ff; figure by author

The political-ideological image of the working woman and sociodemographic aspects of female workforce participation in the GDR

The goal of incorporating women into 'societal production' was pursued
in the GDR as an important step in the realization of equality. Aside
from this continuously propagated aim of the official party of the GDR,
the SED, women also had an important function in the socialist planned
economy as a factor of production. Because of the comparatively low
levels of productivity of the GDR economy, the female labour force was
always required. This situation formed the basis for the development of a
strategy of equality in the realm of politics – for example, on the part of
the politician Inge Lange, a long-time defender of women's interests –
the paternalistic-patriarchal traits of which were not immediately seen.
Thus, the rate of employment among women, high in comparison with
that of western industrialized countries, should not be allowed to obscure
the fact that men and women nevertheless remained social unequals.
The primary reason why the planned economy allowed women to

combine motherhood and participation in the occupational world was because the labour force of women was required.

In terms of the quantitative development of the rate of employment among women and their resulting, relative economic independence, the political, ideological goal really was achieved. The rate of employment among women quickly reached a higher level than in the FRG. However, when considering this 'success', it must not be overlooked that women in the GDR had not only the right but generally also the duty to work. Many women had to join the workforce just for purposes of their own personal financial security or that of their families. Thus, even in 1990, an average of 40 per cent of income in blue- and white-collar households was based on women's earnings (Roloff 1992: 467).

As a result of the efforts to integrate women into the employment system more thoroughly in the early years of the GDR, the birth rate fell. Consequently, an aggressive family policy to increase the birth rate was pursued in the 1970s and 1980s, accompanied by social policy measures aimed at lessening the conflicting pressures of workforce participation and family responsibilities (especially in the form of financial grants and work leaves). The role of women was once again strongly marked by a 'tightrope act between motherhood and the occupational performance pressures of a planned economy' (Nickel 1993: 234). Policy towards women was not conceived in the GDR as a matter of equality; instead, it primarily took place in terms of economic- and social-policy objectives.

Contrary to common beliefs, even in the GDR the employment histories of men and women are more or less clearly distinct from one another. Thus, the occupational history of the majority of women was marked by interruptions. For a quarter of the women, these interruptions lasted longer than a year, and more than 40 per cent interrupted their workforce participation more than once (Engelbrech 1991: 28). During their time of leave, mothers had a right to wage compensation and, thereafter, to re-employment. The legal right to leave covered up to the first 12 months of the infant's life for the first and second child; for the third child, it covered the first 18 months. Single mothers received leave up until the child was 2 years of age if childcare centre capacities could not take on the child. Under these basic conditions, the number of children borne by a woman had only marginal effects on her workforce participation. Thus, in 1988, married women with one child were employed 94 per cent of the time, and even with two children their rate of employment still reached 91 per cent (DIW 1990: 580).

Re-entry into the workforce was made easier by guaranteed work-places and a network of childcare facilities, augmented by generous regulations governing cases of childhood illness (four–thirteen weeks of leave per annum; Institut für Arbeitsmarkt- und Berufsforschung [IAB] 1991b: 140). The availability of childcare centres at all locations was an essential prerequisite for the predictability of female labourers, so that the development of an infrastructure of childcare centres was implemented very early in the GDR (Shaffer 1981: 106). Ultimately, by 1989, 80 per cent of children under 3 years of age were taken care of in a childcare centre for infants (*Kinderkrippe*), 95.1 per cent of children between the ages of 3 and 6 attended a nursery school, and 81.2 per cent of the pupils in lower classes were looked after in a school shelter in the afternoons following school (Nickel 1992: 44).

This comprehensive arrangement of non-family childcare allowed the return of women to the workforce to be more rapid than was the case in the FRG. After age 35, child-related interruptions of female employment hardly played any role whatsoever. Full-time employment – even for women with children – was thus guaranteed by the state. In contrast to its western neighbours, the GDR was not acquainted with the problem of open unemployment. The work week of a full-time employee consisted of 43.75 hours; full-time employed women with two children under the age of 16 worked 40 hours for full pay (Nickel 1992: 47). In 1988, for every 100 married women with two children, 74 were employed full-time and a further seventeen part-time. Though women also performed most part-time work in the GDR, in quantitative terms such work was insignificant. And because it was primarily done by women, part-time work *per se* was never fostered, since it stood in the way of the goal of occupational equality. Moreover, part-time positions were not used to ease the workload of women with dual responsibilities of family and employment, but primarily served in the transition of older women into retirement (Holst and Schupp 1992: 465). Further, part-time work involved much longer hours than in the FRG. Nevertheless, even age groups just below retirement exhibited high rates of employment, since no pre-retirement transitional arrangements existed and employed persons shortly before retirement were given special protection against dismissal. These 'privileges' afforded to mothers on the labour market, however, only increased sex-specific segregation even further.

Sex-specific occupational structures

The rise in the rate of employment among women reflected a rise in the proportion of female employees in almost all areas of the economy. Agriculture and the craft trades represented the only exceptions (Table 8.4). Even though efforts had been intensified in the 1960s to integrate women completely into all sections of the labour market, the relative share of women varied according to branch of industry or occupational group. Attempts were undertaken to increase the number of women employed in industry, leading to a clear rise in the percentage of female industrial-sector employees from 1949 to 1970. However, these efforts – in spite of propaganda to the contrary – became considerably less important after the middle of the 1970s. Indeed, there was even a slight

Table 8.4 Female employees according to economic sector

Sector	1949 %	1970 %	1989 %
Agriculture and forestry	56.8	45.8	37.4
Industry	25.5	42.5	41.0
Craft trades	34.9	40.1	36.7
Construction	9.3	13.6	17.2
Postal and telecommunications services	15.2	36.0	69.0
Commerce	54.0	69.2	71.9
Non-productive fields:	55.4	70.1	72.6
Educational system			77.0
Health system			83.1
Social (welfare) services			91.8

Source After Nickel 1993: 237

reversal in trend, back towards the traditional distribution of sexes across economic sectors.

Nevertheless, different measures for implementing the equality of men and women (for example, wide-ranging educational and advanced-training offensives on behalf of women) resulted in a high level of female educational attainment. Even in the 1960s, as many girls as boys already attended secondary schools. By the middle of the 1970s, equilibrium between the sexes had also almost been reached at the university. Subsequently, in terms of formal qualification structures, hardly any sex-specific differences can still be recognized today among 40-year-olds (Schenk 1992: 34).

Despite the far-reaching convergence in the formal qualification structures of men and women, typical and untypical women's fields of employment were established in the working world. Sex-specific occupational segregation shows that of the approximately 300 occupations for skilled labour (*Facharbeiterberufe*) in the GDR (of which about 30 were closed to women for job-related medical reasons) only about 20 actually played a significant role in the training of girls. There was a high concentration primarily in the service sector (in such skilled occupations as hairdresser, skilled clerical worker and skilled salesperson) and in individual technical occupations in the textile and clothing industries. This concentration even increased in the course of the 1970s and 1980s, and, in 1987, a total of 16 skilled occupations accounted for 62.5 per cent of all females leaving school (Table 8.5). The clearly sex-specific polarization resulted in girls once again increasingly being trained in those occupational fields that already displayed high proportions of women. Conversely, a series of traditionally male, skilled occupations existed in which women accounted for less than 5 per cent of all trainees (for example, plumbers and skilled labourers in metrology and control

Table 8.5 Distribution of females leaving school in occupations of skilled labour[1] in 1987[2]

Skilled worker occupation	Total no.	Women's share: percentage of all workers %	Skilled workers' share: percentage of all female workers %
Skilled sales personnel	8,363	96.5	10.64
Trained personnel for business	7,143	95.4	8.99
Skilled secretarial labour	6,783	99.7	8.84
Cook	5,598	57.6	4.25
Skilled worker[1] in textile technologies	3,099	94.3	3.75
Waiter/waitress	3,274	81.7	3.53
Skilled worker[1] in animal husbandry	3,687	72.5	3.52
Skilled clothing worker	2,565	99.7	3.37
Gardener	2,338	79.1	2.58
Hairdresser	1,911	95.2	2.40
Skilled worker[1] in data processing	1,950	73.3	1.88
Skilled railway custodian	1,473	95.7	1.85
Trained personnel in banking, insurance, finance	1,399	92.1	3.49
Baker	2,130	57.9	1.63
Skilled postal worker	1,309	91.7	1.58

Key
[1] Skilled worker = Facharbeiter
[2] With signed apprenticeship contracts; if Abitur, without occupational training
Source After Schenk 1992: 37

engineering). These sex-typical distribution patterns were not primarily the result of individual choice of apprenticeship or training positions but of the central planning and allocation of such positions. Especially after 1975, there was a sharp drop in the number of technical training positions offered girls. Between 1975 and 1987 the percentage of training positions for girls for the occupation of skilled worker in electronics fell from 49.7 per cent to 20.1 per cent; in the same period, the percentage of girls in training to become skilled workers in metrology and control engineering dropped from 25.9 per cent to 8.4 per cent (Nickel 1992: 41). Thus, at the end of the 1980s, a traditional and narrow spectrum of occupations existed for women in the GDR.

The personnel recruitment policies of factories and enterprises supported sex-specific segmentation mechanisms. In appointment decisions, especially for middle- and high-level positions, enterprises often decided against women in the interest of an effective use of labour force potential since they feared high rates of lost working time due to the comprehensive social-policy measures for mothers. Sex-specific differences were also found in the subsequent use of an enterprise's qualification potential. In this way, women were discriminated against on the internal labour markets within enterprises themselves, and disproportionately often their positions were lower than their level of qualification. Correspondingly, the income distribution also displayed sex-specific differences, with a difference of about 30 per cent in average income existing in 1988 (Schenk 1992: 44). It can be concluded that low-income and low-prestige occupations and branches of industry had the highest percentages of women employees, though 'prestige' did not mean the same in the GDR and the FRG. For example, occupations in the banking and insurance sector were ranked much lower in the GDR than in the FRG in terms of income and prestige and, correspondingly, exhibited much higher percentages of women.

Sex-specific segregation on the labour market was made even worse by restricted opportunities for promotion. Thus, a high level of qualification did not automatically provide women with access to highly demanding positions of responsibility, a gain in occupational status and higher income. In the GDR, jurisdiction over productive and reproductive services continued to be assigned according to sex; that is, even in the socialist planned economy the sexual division of labour was not eliminated.

Sex-specific segregation on the labour market existed on all levels of qualification. Women accounted for about 70 per cent of those with middle-level qualifications (technical college education). The latter were concentrated in the fields of study of medicine, education and economic science, where they accounted for over 80 per cent of the students. Conversely, men dominated in technical courses, especially at schools of engineering. This sex-specific polarization continued to increase during the 1980s. Highly qualified women (with a university degree) were primarily concentrated in the non-productive branches of the economy, such as in the educational system, government administration, or the health system.

THE DEVELOPMENT OF FEMALE WORKPLACE PARTICIPATION IN UNIFIED GERMANY SINCE 1990

At the time of reunification in 1990, the following similarities or differences could be found between the two areas of Germany:

1 Rates of employment among women were markedly higher in the former GDR than in the FRG prior to unification. With an employment rate of almost 90 per cent, the GDR was one of the leading countries in the world, whereas the FRG exhibited a below-average rate, even in comparison with other western industrialized nations. The rate of employment in East Germany among 20–4-year-old women was 66 per cent; among 25–9-year-olds, 79 per cent; and among 30–4-year-olds, 86 per cent. In the pre-unification FRG,

comparable rates of employment were reached only among 20–4-year-old women. All other age cohorts displayed markedly lower rates of employment (for example, only about 60 per cent among 25–9-year-old women) (Brinkmann and Engelbrech 1991). The differences in the upper age cohorts were especially marked. Whereas in the FRG the highest employment rates for women existed in the age cohorts prior to entrance into the family-oriented phase, in the GDR the highest rates were attained in the age groups of 25–45-year-old women.

2 In contrast to the FRG, in the former GDR female workforce participation had been constantly high for years. Due to the chronic scarcity of labour resulting from the low productivity of planned-economic structures, workforce participation practically became a duty for women. There were hardly any regional differences in female workforce participation. By comparison, in the states of the FRG up until unification a slow but gradual increase in the clearly lower rate of female workforce participation could be observed. This was accompanied by a rise in the qualification levels of women, and made it possible for them to pursue occupational-employment and family goals within an integrated concept of female life, with their workforce participation taking the form of part-time work. Moreover, there were still clear regional differences in the workforce participation of women.

3 Part-time work had a different function and a different significance in each of the two parts of Germany. In the former GDR, 25 per cent of all working women worked part-time in 1990; in the FRG, this figure was around 30 per cent (IAB 1991b: 139). Moreover, the average number of hours worked per week by part-time female workers in the GDR (28.7 hours/week) was clearly higher than that of their West German counterparts (20.5 hours/week; Schupp 1991: 267). In the FRG, part-time work was primarily a specific form of employment of mothers who sought either to combine family and paid employment or to re-enter the workforce following the family-oriented phase of their life. Accordingly, the lower and middle age groups among workforce participants displayed the highest proportions of women employed part-time. In contrast, in the GDR, the highest percentage of women working part time was found among the upper age groups, since part-time work apparently played the role of 'initiating' the transition to retirement (DIW 1992: 236).

In the agreement between the Federal Republic of Germany and the German Democratic Republic on the Creation of a United Germany (the so-called Unification Agreement), article 31, paragraph 2, lawmakers faced the task – given the different legal and institutional starting-points in maternal and paternal workforce participation – of designing the legal situation in such a way as to make it possible to

combine family and occupation. Reunification resulted in the quickest
possible implementation in East Germany of the transition from the
planned to the market economy. The process of economic restructuring
with its privatization of nationally owned enterprises had serious consequences for the East German labour market. In contrast to its West
German counterpart, the development of this market took an unfavourable course, especially for women.

When the individual German states are compared, the contrast
between female workforce participation in East and West was still
readily recognizable two years after unification (Figure 8.7). The continuing regional differences in the former West German states are
striking, even though they are clearly smaller than in 1961: the percentage of workforce participants who are women ranges from 37.5 per cent
in Saarland to 44 per cent in Hamburg. All former East German states
reached higher levels of employment, and there was less range of
variation between them (ranging from 45 per cent in Sachsen-Anhalt
to 44.2 per cent in Thuringia). Though some initial signs of convergence
of conditions appeared in the first two years following reunification, even

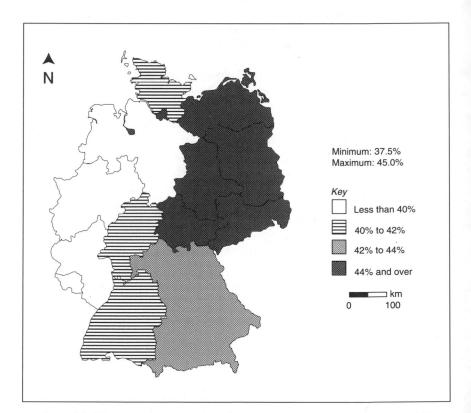

Figure 8.7 Women as percentage of the total workforce, Germany, 1992
Source Statistisches Bundesamt 1993: 135ff; figure by author

in the area of female workforce participation clear regional differences persisted.

Both parts of Germany were hit by a considerable decline in employment in 1993. On the whole, this drop in employment turned out to be weaker in former West Germany than in former East Germany and affected women less than men. At the end of 1993, women accounted for 41 per cent of all workforce participants in former West Germany. In former East Germany, however, women were hit harder by the employment drop. By the end of 1992, they had one third fewer jobs than in 1990, and by November 1992 they accounted for only 43 per cent of all workforce participants as opposed to 47 per cent in November 1989. The elimination of jobs at the beginning of the transformation process especially affected the older age groups among women. Their move into (early) retirement eliminated a great number of part-time jobs, which considerably increased the difference in the part-time employment rates between former East and West Germany (Holst and Schupp 1992). In addition, few of the newly created jobs in East Germany were part-time positions. This meant that women employed part-time were much more affected by the drop in jobs in East Germany than were women employed full-time (DIW 1992: 238).

Thus, since reunification, East German women in particular have been affected by unemployment. Whereas the rate of female unemployment in the area of the former GDR was only 1 per cent in June 1990, by the middle of 1991 it had already reached 14.5 per cent and continued to rise thereafter. Having ultimately reached 21.5 per cent by the middle of 1993, the unemployment rate of women in the 'new' German states (that is, in the former East German states) was not only almost twice as high as that of men, it was also 2.5 times higher than that of women in the 'old' German states (see Table 8.6). Even more marked sex-specific differences existed in terms of long-term unemployment. Due to the concentration of female labour in a small number of occupational areas in the former GDR, just 10 per cent of these areas accounted for 75 per cent of all female unemployment in the area of former East Germany in 1992. In five occupational groups (organizational, administrative and clerical/secretarial occupations; food-related occupations; commodity-sales personnel; clothing and textile occupations; and cleaning professions), women made up over 90 per cent of the unemployed (Jasper 1993: 115). Finally, some areas of work that had previously been largely dominated by women (such as the banking, credit and insurance industry) were increasingly taken over by men after reunification. The number of male employees in the tertiary sector as a whole rose in the first year following unification, while that of women dropped (Holst and Schupp 1992: 470). This development led some to characterize the formerly employed East German women as the losers in the process of German unification, especially since currently women have few prospects of escaping unemployment.

Table 8.6 Development of the rates of unemployment in the 'old' and 'new' states according to sex[1]

Year	'Old' states		'New' states	
	Men	*Women*	*Men*	*Women*
1991	5.7	7.2	9.8	14.5
1992	6.2	7.4	10.4	20.1
1993	8.1	8.8	11.0	21.5

Key
[1] Figures refer to the middle of each year.
Source After Engelbrech 1994: 15

The secondary sector was particularly hard hit by job elimination. In the textile industry, for instance, in which two thirds of those employed were women, one of every three jobs was lost between September 1989 and the end of 1990. This loss was especially felt in the south of former East Germany. In the chemical industry (around 40 per cent of the workforce of which was female), about every second job was lost, with Sachsen-Anhalt losing an especially large number of these positions.

PROSPECTS

Given the differences that still exist and a convergence that is only slowly taking place, various authors assume that the short term will not bring about a complete convergence in the state of female workforce participation in the 'old' and 'new' states (for example, Engelbrech 1991: 21). This is true even though changed living conditions, such as in the area of reproductive behaviour, have already resulted in recognizable trends in the direction of West German conditions (for example, Roloff 1992: 474). The gradually occurring elimination of differences will tend, in the estimation of all authors, to move the 'image of the woman in line with a traditional polarizing pattern' (Nickel 1993: 255) and thus largely eradicate the image of the woman found in the former GDR. However, the previous model of the working woman will continue to have its place in the moral conceptions of East Germans for some time to come.

According to a scenario envisioned by Stooss and Weidig (1990), by 2010 there will be a further increase in workforce participation in the 'old' states of the FRG in conjunction with a continuing tertiarization of the economy. Since especially those parts of the service sector in which women display high rates of employment (commerce, general services) will lose in relative importance, the only way that the rate of employment among women can continue to rise is if the current spectrum of female workforce participation is extended to include those service-sector branches that will prosper in the future (especially enterprise-oriented services). The potential for a further increase in female workforce participation continues to reside above all with women with children (Vogel 1993: 358).

In the 'new' German states, the transition from the planned to the market economy will be accompanied by a deep sectoral-related transformation of the economy. This will continue in the coming years and will take place more rapidly than in the 'old' German states. Thus, continued strong losses of employment are expected for those with low qualifications, especially in the typically female occupations. Here, priority must be given to the task of adjusting the future educational profile of women to the clearly raised qualification requirements. Just as in the 'old' German states, one finds in the area of the former GDR that the greater the participation in education, the greater the tendency to work (Institut für angewandte Sozialwissenschaft 1991). In general though, this trend is connected with a much broader tendency among currently non-working East German women to seek work than is found among their counterparts in former West Germany (DIW 1992: 240).

The greater orientation of East German women towards workforce participation, often proven and still markedly higher today, can basically be explained in terms of economic needs. East German women continue to be important in producing a large part of household income, so that we can assume that the extent of their workforce participation will not move towards current West German levels. Nevertheless, under the changed economic and societal conditions now prevailing, female workforce participation will also not reattain the levels it enjoyed during the existence of the GDR. Since at the same time the workforce participation of middle-aged women is expected to rise in West Germany, it can be assumed that the future rates of employment among women in East and West Germany will tend to converge on a level somewhere between their respective current levels.

REFERENCES

Amtliche Nachrichten der Bundesanstalt für Arbeit (1989) No. 6, Nuremberg: ANBA.
—— (1991) No. 1, Nuremberg: ANBA.
Beilner, H. (1971) 'Emanzipation der bayerischen Lehrerinnen: aufgezeigt an der Arbeit des Bayerischen Lehrerinnenvereins (1898–1933)', *Miscellanea Bavaria Moncensia* 40, Munich: Woelfle.
Brinkmann, C. and Engelbrech, G. (1991) 'Erwerbsbeteiligung und Aspekte der Erwerbstätigkeit von Frauen in der ehemaligen DDR', *Beiträge zur Arbeitsmarkt und Berufsforschung* 143, Nuremberg.
Bundesanstalt für Arbeit (1993) *Teilzeitbeschäftigte in Westdeutschland*, imu 991190, Nuremberg.
Bungardt, K. (1959) *Die Odyssee der Lehrerschaft. Sozialgeschichte eines Standes (ein Versuch)*, Frankfurt: Kern & Birner.
Deutsches Institut für Wirtschaftsforschung (DIW) (1990) 'Vereintes Deutschland: geteilte Frauengesellschaft. Erwerbsbeteiligung und Kinderzahl in beiden Teilen Deutschlands', *DIW-Wochenbericht* 41: 575–82.
—— (1992) 'Umbruch am ostdeutschen Arbeitsmarkt benachteiligt auch weiterhin die erwerbstätigen Frauen: dennoch anhaltend hohe Berufsorientierung', *DIW-Wochenbericht* 18/92: 235–41.

Engelbrech, G. (1987) 'Erwerbsverhalten und Berufsverlauf von Frauen: Ergebnisse neuerer Forschungen im Uberblick', *Mitteilungen aus der Arbeitsmarkt- und Berufsforschung* 2: 181–96.

—— (1991) 'Der Arbeitsmarkt für Frauen in den alten und neuen Bundesländern', *Beiträge zur Arbeitsmarkt- und Berufsforschung* 167 (Nuremberg): 20–32.

—— (1994) 'Frauen: nur Reservearmee für den Arbeitsmarkt?', *Die Frau in unserer Zeit* 1: 14–23.

Helberger, C. (1988) 'Frauenerwerbstätigkeit und die Entwicklung der sozialen Sicherungssysteme im internationalen Vergleich', *Zeitschrift für Sozialreform* 39: 735–49.

Holst, E. and Schupp, J. (1992) 'Frauenerwerbstätigkeit', in Statistisches Bundesamt (Hrsg.) (1992) *Datenreport 1992. Zahlen und Fakten über die Bundesrepublik Deutschland. Schriftenreihe der Bundeszentrale für politische Bildung* 309: 463–70.

Institut für angewandte Sozialwissenschaft (INFAS) (1991) 'Frauen in den neuen Bundesländern im Prozeß der deutschen Einheit', in *Materialien zur Frauenpolitik*, 11, Bad Godesberg: Bundesministerium für Frauen und Jugend.

Institut für Arbeitsmarkt- und Berufsforschung (IAB)(1991a) 'Qualifikation und Arbeitslosigkeit in West- und Ostdeutschland', in *IAB-Kurzbericht*, No. 12, 12, Nuremberg.

—— (1991b) 'Erwerbsbeteiligung und Aspekte der Erwerbstätigkeit von Frauen in der ehemaligen DDR', in *IAB-Kurzbericht*, 42, 12, Nuremberg.

Jasper, G. (1993) 'Zur Krise der Frauenerwerbstätigkeit in den neuen Bundesländern', in K. Hausen and G. Krell (eds) *Frauenerwerbsarbeit. Forschungen zu Geschichte und Gegenwart*, Munich: Hampp.

Knauth, B. (1992) 'Frauenerwerbsbeteiligung in den Staaten der Europäischen Gemeinschaft', *Acta Demographica*: 7–25.

Maier, F. (1993) 'Zwischen Arbeitsmarkt und Familie: Frauenarbeit in den alten Bundesländern', in G. Helwig and H. M. Nickel (eds) *Frauen in Deutschland 1945–1992*, Berlin: Akademie-Verlag.

Meusburger, P. and Schmude J. (1991) 'The relationship between community size, female employment rates and the educational level of the female labour force', *International Geographical Union Study Group on Gender and Geography Working Paper* no. 12.

Myrdal, A. and Klein, V. (1956) *Woman's Two Roles: Home and Work*, London: Routledge & Kegan Paul.

Nickel, H. M. (1992) 'Frauenarbeit in den neuen Bundesländern: Rück- und Ausblick', *Berliner Journal für Soziologie* 2: 39–48.

—— (1993) 'Mitgestalterinnen des Sozialismus: Frauenarbeit in der DDR', in G. Helwig and H. M. Nickel (eds) *Frauen in Deutschland 1945–1992*, Berlin: Akademie-Verlag.

Recum, H. v. (1955) 'Nachwuchsprobleme des Volksschullehrerberufes in Schleswig-Holstein', *Soziale Welt* 2/3: 153–65.

Roloff, J. (1992) 'Zu Problemen der Erwerbsbeteiligung der Frauen in den neuen Bundesländern', *Zeitschrift für Bevölkerungsgeographie* 4: 465–75.

Schenk, S. (1992) 'Qualifikationsstruktur und Qualifikationsbedarf erwerbtätiger Frauen in den neuen Bundesländern', *Beitrage zue Arbeitsmarkt- und Berufsforschung*, 167 (Nuremberg): 33–48.

Schmude, J. (1988) 'Die Feminisierung des Lehrberufs an öffentlichen allgemeinbildenden Schulen in Baden-Württemberg', in *Heidelberger Geographische Arbeiten*, 87, Heidelberg: Selbstverlag des Geographischen Instituts der Universität Heidelberg.

Schupp, J. (1991) 'Teilzeitarbeit in der DDR und in der Bundesrepublik Deutschland', in Projektgruppe Das Sozio-ökonomische Panel (ed.) *Lebenslagen im Wandel. Basisdaten und -analysen zur Entwicklung in den neuen Bundesländern*, Frankfurt/New York: Campus.

Shaffer, H. G. (1981) *Women in the Two Germanies. A Comparative Study of a Socialist and a Non-Socialist Society*, New York: Pergamon Press.

Statistisches Bundesamt (1964) *Statistisches Jahrbuch 1964 für die Bundesrepublik Deutschland*, Stuttgart/Mainz: Kohlhammer.

—— (1971) *Statistisches Jahrbuch 1971 für die Bundesrepublik Deutschland*, Stuttgart: Kohlhammer.

—— (1987) *Statistisches Jahrbuch 1987 für die Bundesrepublik Deutschland*, Stuttgart/Mainz: Kohlhammer.

—— (1989) *Statistisches Jahrbuch 1989 für die Bundesrepublik Deutschland*, Stuttgart: Kohlhammer.

—— (1991) *Statistisches Jahrbuch 1990 für die Bundesrepublik Deutschland*, Wiesbaden: Metzler-Poeschl.

—— (1993) *Statistisches Jahrbuch 1993 für die Bundesrepublik Deutschland*, Wiesbaden: Metzler-Poeschl.

Stooss, F. and Weidig, I. (1990) 'Der Wandel der Tätigkeitsfelder und -profile bis zum Jahr 2010', *Mitteilungen aus der Arbeitsmarkt- und Berufsforschung* 1: 34–51.

Twellmann, M. (1972) 'Die deutsche Frauenbewegung: Ihre Anfänge und erste Entwicklung, 1843–1889', *Marburger Abhandlungen zur politischen Wissenschaft*, 17, 1, Meisenheim am Glan.

Vogel, C. (1993) 'Ausgewählte Strukturdaten zur Frauenerwerbstätigkeit', *Baden-Württemberg in Wort und Zahl* 9: 357–9.

ZVS (ed.) (1958) *Statistisches Jahrbuch der DDR 1957*. East Berlin: ZVS.

—— (1971) *Statistisches Jahrbuch der DDR 1971*, East Berlin: ZVS.

—— (1988) *Statistisches Jahrbuch der DDR 1988*, East Berlin: ZVS.

9

THE POLITICS OF CULTURAL IDENTITY: THAI WOMEN IN GERMANY

Eva Humbeck

While in theory all residents are expected to participate in the political and economic improvements the European Union (EU) promises, the likelihood of benefits for women has been questioned (see Kofman and Sales, Vaiou and Valentine in this book). Even more questionable is the benefit for immigrant women from non-European countries. Some of these women are in Europe illegally and lack protection from the law; for others, their legal status is often tied to their marital status. The restrictions on their lives may even be increasing as a closer-knit EU is in the process of adopting tighter restrictions in policies to regulate immigration from outside Europe, which will grant or refuse residency rights in all or none of the European countries bound by the Schengen Agreement. The fear of a 'Fortress Europe' which eases life for Europeans while making it harder for individuals from outside seems to be justified.

Central to debates about the place of immigrants within the EU is the issue of cultural identity and marginalization or integration of people perceived as 'other'. The substantial number of women from Third World countries now living in the EU clearly fall within this category. Particularly vulnerable to exploitation are young women from Asia and Latin America who have migrated independently of their families of origin. These migrants find themselves in a new environment where they may not be able to speak the language, may have restricted employment opportunities and must forge new support networks. Yet they are also women who have shown initiative to migrate to Europe with the objective of improving their lives. They are now confronted with the challenges of forming new identities in an alien environment.

In this chapter I will be concerned with one such group of women – Thai women who have married German men. Reports indicate that by 1990 approximately 1,000 such marriages were taking place each year, in addition to roughly 2,000 marriages per year by German men to women from the Philippines, the Caribbean, Latin American and African countries (Statistisches Bundesamt 1989). While I am most interested in the motives of these Thai women and the ways in which they are shaping their lives and identities, I will also examine how German social and political institutions are positioning themselves in relation to the Thai women, thereby revealing their own expectations about the place which 'other' women should take in German life. I do not suggest that the experiences of these Thai women are representative of all Third World migrant women in EU countries. Women who are alone, or those concentrated in domestic employment, for example, will have different experiences (see Hillman forthcoming). So will the women who are caught up in the sex trade (Morokvasic 1993). The case, however, permits me to explore how issues of identity are framed by both immigrant women and native Germans, by community agencies and by the state, and to show how gender, sexuality and 'race' come together in a transnational context. My discussion is drawn from a larger research project (Humbeck 1994) that examined the situation of these women in Thailand and in Germany. In that project I combined an array of methods to obtain information: interviews with Thai informants and field observations in Thailand; interviews with representatives of German governmental agencies and non-governmental organizations concerned with the welfare of Thai women in Germany and a review of documents they publish; participant observation at social events involving the Thai wives, together with unstructured group discussions and in-depth interviews with selected women; and a questionnaire survey directed to German husbands and Thai wives.[1]

For several reasons, Germany presents an important context within the EU for studying immigrant issues. First, in absolute terms, Germany has been receiving larger numbers of immigrants and refugees than other member states of the EU and it is a destination in demand. During the 1980s, the number of applicants for political (and economic) asylum from eastern Europe as well as from non-European countries rose dramatically, from already twice the average number of applications to EU countries in 1987 to an unprecedented 256,000 in 1991, far outstripping applications to the UK, the second-ranked country, which received only 57,700 (as cited in Heisler and Layton-Henry 1993: 151). A high proportion of applications are unsuccessful, but the percentage of all residents of Germany who are of foreign origin now reaches about 8.2 per cent (Heisler and Layton-Henry 1993). Second, Germany is one country where the recent wave of immigration has prompted numerous anti-foreigner demonstrations by right-wing political groups. It thus becomes of heightened importance to gain insights into the ways various actors in Germany society, in addition to the immigrant women themselves, are constructing the identities of immigrants.

Much has already been written about immigrant groups in Germany, including some research studies on women, particularly those from Turkey and Yugoslavia, and there is growing attention to the theme of women and international migration in general (see, for example,

Abadan-Unat 1977; Aziz 1992; Davis and Heyl 1986; Morokvasic 1984, 1991a, 1991b, 1993; Munscher 1984; Phizacklea 1983; Simon and Brettell 1986). Researchers are exploring women's labour force participation, their susceptibility to unemployment and the implications of migration for family and gender relations. More recently, the situation of eastern European women, including their position in the sex and marriage trade, has attracted researchers (Morokvasic 1993). By comparison, attention to Thai women has largely been in the popular press, where it focuses on the sex trade, and they have not been the subject of much scholarly research. Neither has the issue of women in bi-cultural marriages been significantly explored.

Before taking up the case of Thai women married to German men, I wish to indicate the diversity of Thai women in Germany. Available records show that in 1990 about 14,000 Thai women were living in Germany, plus an estimated 5,000 residing there illegally. An estimated 10 per cent of these have entered on three-month tourist visas and work illegally in the restaurant and entertainment businesses, more specifically in prostitution. These women are usually brought in by agencies which organize their work, living and travel arrangements at high costs. Often the women are lured into this situation under false pretences of employment in the hotel or restaurant industry with promises of making a lot of money, only to find themselves trapped in organized prostitution with little hope of earning enough to repay the costs of their trip to Germany. They are generally distributed throughout the country on a rotating schedule to prevent detection and expulsion by the authorities. If they should be able to escape their employers, they seldom turn to the police for fear of being deported and losing the money they have invested (Morokvasic 1993). They are completely dependent on others, have no legal rights, do not intend to stay in Germany for a prolonged time period and have minimal contact with Germans and German culture.

By estimate, another 5–10 per cent fall into the categories of students and/or highly educated professional women, often from privileged Thai families. Some of these women are married to Thais, others are single and some are married to Germans. This group usually finds a supportive social and working environment (that is, husband, family, legal work or study visa). They are articulate, often very well versed in the German language, aware of their opportunities and alternatives in both countries, and they master and apply the skills of switching from Thai to German cultural behaviours sufficiently to be effective in each context.

The majority, or an estimated 80–5 per cent of Thai women in Germany, fall into a third category. These women arrive in Germany by arrangements made through a marriage agency or private contacts with the intention of marrying a German man. They usually enter on a three-month tourist visa after being selected by an interested German client via catalogue or video, or after meeting their prospective husband in Thailand. These women have been described as generally having little formal education and speaking no German. Often, they cannot even read or write Thai very well (AGISRA 1990). On the average, they are between 18 and 35 years of age. Most come to Germany directly from Bangkok but may originally have lived in the economically depressed provinces of the north and the north-east of Thailand. Some have worked in prostitution or related occupations in Bangkok, some have held jobs in restaurants and hotels, in manufacturing, or in sales. Given a tourist visa only they intend to get married to a German husband as soon as possible. If their visa expires before they are selected, most will try again and take out another loan to pay for an opportunity to return to Germany. Individual factors, such as length of stay, social environment in Germany, support from the husband and so on, vary for these women. It is for this group that politics of identity formation are most problematic if they wish both to establish a long-term life in Germany and also to maintain the option of returning to Thailand. They are the central focus of my research.

INTEGRATION: THE OFFICIAL VIEW

In 1990 (then West) Germany had a population of 61 million citizens and almost 5 million foreign nationals and was dealing with an increasing stream of political refugees seeking asylum, rising from about 60,000 in 1987 to almost 200,000 in 1990 (Tichy 1993). In addition, a quickly accelerating number of refugees of German heritage from former socialist countries of eastern Europe poured into Germany in 1989 and 1990 (Tichy 1993). This poignant situation became even more apparent when East Germany opened its frontier to the West, and a flood of East German migrants had to be supplied with shelter, work and services. Germany, its authorities as well as many of its citizens, had long proposed a policy of integration rather than promoting a multicultural society as demanded by the Green Party and various activist groups around the country (*Die Grünen* 1990). Today, complete integration is still encouraged and promoted under the assumption that it will keep inter-racial problems at bay and preserve the status quo. According to the German government, it is intentionally sending such a strong message to discourage future immigrants and to be able to restrict the number of political refugees to the minimum. Officials and citizens feel this is necessary in order to prevent Germany from being overrun by economically motivated immigrants. During times of economic recession in the late 1980s and early 1990s, problems arose in providing shelter, jobs and social services. Under the perceived pressure from its constituency, the Christian-Democratic-led government thought it necessary to curb lenient immigration and asylum laws with effect from 1991.

The pressure exerted by the authorities and by society on immigrants is very powerful and inescapable. The dominant view taken by society and authorities is still that integration should be the desired goal for immigrants to Germany. Since the experience of German unification in particular, the discussion about the extent and effects of cultural adjustment of immigrants has become more heated and personal (Tichy 1993). Often, citizens or politicians do not realize that migrant adaptation is a process of change for the host society as well as the migrant population. The official position has hardened with the tightening of immigration and asylum laws in 1991, even while the number of immigrants arriving has decreased. Despite efforts by the Green Party and other local political groups who have pushed for the institution of a multicultural nation, no major progress has been made towards that goal other than on a regional level, for example, by the Hessian state parliament.

Germany does not consider itself an immigrant country, and its laws are to protect its citizens first; that is, the husbands, the marriage agencies, or the employers. Thus Thai women have found it difficult to pursue legal suits against physical and psychological abuse or child custody and residency claims and their efforts have often ended unsuccessfully, especially after divorce or separation from their husbands. Thai women are only partially protected by German law even when married to a German. They stand to lose their residency in case of death of spouse or divorce within the first three years of marriage, or five years, respectively, if the couple has children. German courts generally grant child custody to the German father who is thought to be able to provide a better economic future for the child. The Thai mother, on the other hand, upon losing her residency, can then be forced to leave the country within weeks.[2]

The German government has increased its efforts during the past years to control illegal businesses facilitating immigration of Thai women for the purpose of prostitution. Because of the activities of human rights organizations and women's groups in Germany as well as on an international level, the situation of Thai women's migration to Germany has been widely publicized. Debates about international traffic in women and the protection for women from these practices in Germany have been initiated in parliament by women representatives from all political parties and several reports have been published on the situation through government initiative (Beauftragte der Bundesregierung 1989). Although many politicians seem to agree on the unequal, exploitive and often inhuman treatment of these women, the legislature has been slow, even reluctant, to change the legal status in order to grant them more protection.

COMMUNITY ORGANIZATIONS: COMPETING VISIONS

The cultural identities and experiences of immigrant Thai women are strongly influenced by the fact that most of them live in isolation from other Thais. Outside cities like Frankfurt, Berlin and Hamburg, which are known for a higher than average percentage of foreigners, no clustering or large population of Thais occur. Therefore, most of the social contacts available to Thai women are with individual Germans, German groups, and German organizations. Here I will examine the contrasting approaches of non-governmental organizations (NGOs) towards Thai women.

Several important and influential NGOs take an interest in aiding the process of adaptation of Thai women in Germany. They can be split into two main interest groups – activist NGOs and humanitarian-religious NGOs. Activist NGOs have their origins in the political activist scene. Their involvement is often based on the social criticism of the student movement and/or feminist ideology. Their ideological positions and images often result in their being marginalized by the public, authorities and institutions whose perceptions they are trying to change. One of the best-known groups in this category is AGISRA, an organization working against international sexist and racial exploitation, in particular of women. Another is IAF (1991), an initiative to help people in bi-cultural relationships and families. Both organizations have their headquarters in Frankfurt but also maintain local chapters in most major cities. In addition, several local women's groups and feminist initiatives have established women's shelters, such as the Ban Ying in Berlin.

Members of these NGOs are usually German women who volunteer or work as part-time employees. The non-profit structure of these agencies allows eligibility for government funds and grants. The workers are highly motivated and active in organizing media coverage, helping women in emergencies, organizing protests, offering German language courses and regular informal meetings for foreign and German women, as well as compiling libraries of grey literature on related topics. Their actual contact with Thai women is often limited, however. No all-Thai activist group existed in 1990, although some Thai women were involved as co-workers and counsellors and translators in activist NGOs where very few German women speak Thai. The structure of these organizations is often quite hierarchical and, as many Thai women complained, their German co-workers remain dominant in the decision-making processes and often only delegate work to them.

Activist NGOs generally embrace the concept of a multicultural society. They promote self-reliance, independence and cultural awareness in their programmes of activities. They pursue these goals by political action, educational and counselling services, support groups and so on. Because they also wish to support the options of Thai women to return home, they generally favour a cultural model of 'code switching' whereby the women can maintain identity in both cultures and employ behaviours from each at will over a model of integration into German culture. Their effectiveness in actually reaching Thai women, however, is quite limited.

The humanitarian-religious NGOs include religious groups such as the Ökumenische Thai Gruppe in Frankfurt and a missionary organization,

Saisamphan, but also the Thai Student Club, the Thai Club and the Deutsch-Thailändische Gesellschaft. These groups have very different backgrounds and modes of operation, ranging from very formal clubs to informal gatherings. Their approach differs somewhat from that of the other NGOs as they work with a less politically but rather more socially oriented outlook. Their goals with respect to issues of cultural identity are to support the development of personal independence and self-reliance for Thai women by fostering their knowledge about existing social systems. They also work through counselling, education and social functions, but in a less formal way than the activist NGOs. Some of these groups also favour the cultural 'code-switching' model, whereas other more conservative groups envision integration into German culture as the ideal.

Various means are employed by these religious-humanitarian groups to work towards their objectives, which reflect their conceptions not only of culture but also of gender. They provide education and psychological and communication counselling, and focus on establishing social net-working, support groups, contacts with friends, practical advice on food, shopping and socializing. To some extent they undertake political activism on behalf of the women and are sensitive to the social pressures, prejudice, racism and sexism that the Thai women confront. Their effectiveness, in establishing social contacts and providing useful information can be marked, though the German groups seem to be more helpful in providing support and facilitating understanding of the adaptation process for the German husbands, friends and other Germans than for the Thai wives. Thai women prefer to rely on Thai organizations which can provide them with more than advice but also serve as a point of cultural contact and strengthen their sense of Thai identity.

A number of problems arise in the functioning of both the activist and religious-humanitarian agencies. First is their lack of effective communication with official government agencies, as well as a sense of rivalry between the governmental agencies and the NGOs, and among the NGOs themselves. Second are differences in expectations about cultural models and desired outcomes, so that they sometimes seem to be working towards conflicting and even contradictory goals. Third is their lack of sensitivity towards the goals of the Thai women themselves, with their German ethnocentricity posing an unrecognized problem among agency workers. Thai women who have organized their own support groups either fight this form of paternalism vehemently or, more commonly, react by simply ignoring the efforts made by German groups and by relying on their own systems of support. Another aspect of existing ethnocentricity is that the many organizations offering help operate from a basic western humanitarian point of view. They often do not realize that some of their proposed solutions might not be culturally acceptable to Thai women by, for example, causing them to lose face or, more importantly, denying their Thai cultural identity.

THE POLITICS OF PERSONAL LIFE: MARRIAGE AND FAMILY, SECURITY AND IDENTITY

A great deal of the research into the experience of immigrants in 'host' societies has concerned itself with migrant communities that have been numerically male-dominated or in which women are examined as wives and daughters of immigrant men. The women are thus seen within the context of their own ethnic groups. But for the Thai women, initial contact with the new culture is at a personal rather than group level. Thus to understand the developing identities of the women, it is important to consider as well the attitudes, expectations and behaviour of their German husbands and the relationships within marriage. I will begin by looking at the background and motivations for marriage of the German men and the Thai women. Then I will address aspects of the marriage relationship.

The German husbands and their motives for marrying Thai women

It is commonly believed that the German men who seek Thai (and other foreign-born) wives are older on average than men marrying German women, that a majority are divorced or widowed, and that they are less well educated than German men in general. Although the number of husbands from whom I gained information is not large, there is no reason to think that they are unrepresentative of German men married to Thai women. I found them to be a diverse group. They ranged in age from 25 to 65 years, ten being 41 years and older, and eleven between 25 and 40 years old. The educational level of these men also varied. Six men have a basic, primary education, nine have completed a secondary or technical education, and six hold a university degree. Contrary to findings by Latza (1987) and TPF (Tübinger Projectgruppe Frauenhandel 1989), a majority do not have a blue-collar, working-class background but work in white-collar, service-related occupations (N = 13). Only four work in low-paying, blue-collar positions, two are self-employed, one works in academia, and one is retired. The most common annual income level of 50,000 to 70,000 DM per household (N = 8) also points towards a middle-class rather than a low-income economic background. This is substantiated further by the fact that eight couples own their homes or apartments. Ten couples rent apartments, and two live with relatives.

While the literature often characterizes the men as divorced or widowed with children from former marriages (Katholische Frauengemeinschaft

Deutschlands [KFD] 1981), my findings cannot fully support that statement. For four men this was their first marriage, and only three men admitted that they had been married before; however, because a large number of participants refused to answer this question (N = 14), there might be reason to believe that more men were previously married to German women. Only two reported children from a prior marriage. Since my questionnaire was long, and probed for private information, many men appeared to be suspicious, did not provide detailed information and avoided answering some questions.

The literature on sex-tourism suggests that German men in seeking a Thai wife are motivated by desires for an exotic and docile partner (Cohen 1982, 1986; Institut für Interdisziplinare Sexualforschung 1988; Lipka 1987; Latza 1987). In addition, it has been reported that German men who marry Thai women do so because of their lack of success with German women (Ruenkaew 1990). Among my informants, only one reported finding his wife through a newspaper advertisement, marriages more commonly being arranged through friends or resulting from meeting Thai women during vacations or work-related visits to Thailand. Almost half the men indicated that the relationship was not planned and was a result of fate, chance, or 'falling in love'. Some did suggest that Thai women are 'pretty and sweet' and make better wives, and that they found German women too liberal, egocentric, pushy and emancipated. A number indicated that they wanted a wife who would be happy to stay at home, care for husband and children, and accept her husband's decisions.

The Thai women and their motives for marrying German men

Marriage agencies in Thailand prepare videotapes which list prospective brides in catalogue fashion. One agency reported that the typical Thai woman interested in marriage to a foreigner is between 20 and 29 years of age, has some knowledge of English – from little to fair – and is single without children. About one fifth of the women included in their catalogue of over 200 women have been divorced or widowed, and half of these report having no children. The age range in the catalogue is 17–43, most women being between 20 and 35 years of age. About half indicate they have elementary or high school education, the others have some college or technical schooling or are university graduates. They list a variety of occupations – accountant, tour guide, student, receptionist, designer, teacher, restaurant owner, bank clerk and so on.

The women I interviewed in Germany displayed some of the same characteristics. They were generally younger than their husbands but ranged in age from 18 to 50 at the time I met them, with slightly more than half in their thirties. About one third of the women stated that they had received a primary education, another third indicated a secondary education or technical education, and the remainder reported college or university education. In this, they resemble the women in the video catalogue, although they are considerably better educated than the women described in other research as being poor, rural and uneducated (Phongpaichit 1982; AGISRA 1990). We should acknowledge, however, that some of the 'colleges' are beauty schools and similar professional enterprises. Nevertheless, my data suggest that the women are relatively comparable with the men in terms of their level of education. All of the women I interviewed had worked outside the home in Thailand – about one third in sales, office, or other services, about one quarter in hotels, bars, or tourist-related services, though none indicated a connection to the sex-tourism industry. Given that Thai women suffer from being stereotyped as prostitutes in Germany, it would not be likely that they would report such a connection if it existed. Only three of the women had worked as farmers or unskilled labourers.

According to the literature, the majority of the Thai women who decide to marry a foreigner do so for one main reason – to escape the economic pressure imposed on many families in the rural areas of Thailand (Lipka 1987; KFD 1981). My data confirm the importance of economic motivations, but present a more complex view. The obligation to help their families economically is traditional, and a very strong one for any Thai. Most of the women looking for marriage in Bangkok come originally from the rural areas of the north and north-east. Their families are poor, often landless and indebted to creditors. In Buddhist tradition, daughters especially feel compelled to help their families financially as well as to take care of ageing parents. If they cannot find employment near the villages, which is usually the case in the north and north-east, they are forced to migrate to the larger cities including Bangkok (Phongpaichit 1982).

Job opportunities in Bangkok for unskilled labour are relatively scarce, and employment in the sex-tourism sector provides a lucrative prospect, as well as the opportunity to meet foreigners. While some women are tricked into abusive relationships with foreigners and into prostitution in Germany by the false promises of recruiting agencies, many take the initiative themselves to contract marriages which seem to provide a way out of prostitution or a means of avoiding it. While not all women who marry Germans have necessarily worked in the sex-tourism sector or have had contacts with westerners, many view the opportunity to marry a 'farang' as a sound economic investment (AGISRA 1990; Lipka 1987).

A second important motivation is desire to enhance their social status; marriage to a foreigner is seen as a way to attain this goal (Lipka 1987). Germans are thought to be relatively wealthy since they obviously can afford to take long vacations and spend a lot of money in Thailand. Germany is only vaguely known but assumed to be a modern, leading industrialized nation where many amenities are common, and thought to have a great need for cheap labour. The perception prevails that all of these circumstances render it a better place to make a living than Thailand. Life in Germany is also rumoured to be easy for wives because women with affluent husbands do not have to engage in gainful employment if they do not want to, but instead can afford to stay at home, a privilege of the rich in Thailand. Female tourists from Germany are seen to wear nice clothes, to be treated well by their husbands, and to own many expensive things. Any children of mixed ethnic background would have Eurasian features, which in Thailand are considered beautiful and superior to Asian looks.

Yet another reason for marrying a German man is his reputation. Germans are seen as family-oriented and as dependable husbands who treat their wives well (Lipka 1987). Unlike Thai men, they are assumed not to spend all their money drinking and gambling with their friends, or to disappear and leave their wives and families without support. According to Lipka (1987), and supported by my survey, the majority of women believe that Thai men are unreliable, do not take care of their wives and family, and do not spend much time with them. Thus Thai women see advantages in marrying a German.

I asked women how each had met her husband and why she had decided to marry him. Eight women said they met in Thailand while the man was on vacation; four met through work contacts in Thailand or Germany; two women went through an official marriage agency, and six said that friends or relatives already married to Germans arranged the meeting in Germany. These responses indicate that personal contact is preferred and probably more effective than the anonymous arrangement by marriage agencies. The women also usually emphasized that chance played a role instead of arrangement, and that others, friends or family, who had already participated encouraged them.

The reasons they gave for selecting a German spouse differed interestingly from those the men gave for their selection. About half of the women said that they married their husband because he was a good person and they felt sympathy for him. About a third stated that German men have more respect for their wives and family and treat them well. Two women thought it was fate that they met and married; but not one woman cited love as a reason. The women thus differ from their husbands who commonly expressed concepts of 'fate' and 'love'. Here we may be seeing a significant cultural difference between expressions of western romantic concepts and Thai values. The women spoke of their husbands in terms of respect, loyalty, sympathy and thankfulness. These values are very important in Thai culture and are reflected in some of the women's responses:

> Met my husband in Singapore on vacation; my friend played matchmaker; we got married two years later. My husband is a good husband and a future father, but not because he is German. (questionnaire A1)

In 1987 I and my friend had three weeks' summer vacation and my friend invited me to visit her family. They have a restaurant abroad (Germany). We visited them together. At that place I met my husband. At that time we both (my husband and I) had been working hard, so we (were ready to meet) each other; we talked like friends. After that my husband stayed in contact with me after I returned to Thailand. (questionnaire A23)

I came with my husband to Germany to escape poverty. I am from a village, a person who does not know much. (questionnaire A2)

He is a very good person, good enough to choose him as a husband. I am lucky to have him as a husband. (questionnaire A22)

Marriage and family, security and identity

Almost unanimously, the Thai women who responded to my survey questionnaire reported that family was the most important aspect of their lives. Here they differed from their husbands, only about half of whom placed family first. Probing these evaluations, I discovered that the men more often than the women placed high value on emotions (as they had in discussing their marriages) – love, harmony, faithfulness – whereas for the women financial security, in addition to their families in Germany and their families in Thailand, loomed large, though they also valued faithfulness, harmony and being treated fairly. The men also placed a higher value on their own independence and on social life and friendship than the women did. These findings suggest a pragmatic approach to life among the women and a centring of life around the family. These experiences and values were expressed in interviews as well as in the questionnaire responses and it became clear during the study that many of the decisions the women made were based on consideration for their children in Germany; about one third of the women also took into account the well-being of their children in Thailand who were being raised by relatives. The women also reported that most of their activities were centred around the home and they mentioned loneliness and isolation as problems. Especially in the first few years of marriage they are dependent on their husbands for support and social contact and the husband is in quite a powerful position, given his economic resources and residence within his own cultural, social and linguistic environment. Communication within the home is also predominantly in the husband's language.

The women did not present themselves as victims, however. While not appearing as emotionally involved in their marriages as the men, they expressed sympathetic feelings for their husbands, were supportive and willing to live with them. Slightly more than half were working outside the home or studying (either in language classes or towards career goals) and did not report resistance to their activities by their husbands. More than two thirds of them indicated that most decisions in the marriage were taken jointly. Generally, the women rated themselves as better off than when living in Thailand. They seemed to see the marriage more as a fortunate economic opportunity to better their lives and those of their families in Thailand.

Nevertheless, only five of the women indicated that they saw no problems in being married to a German and living in Germany; about one third saw it as 'acceptable' and about a third indicated that they would like to leave but stayed because of their husband and children. Almost all expressed the wish to return to Thailand at a later point, however, after the children had grown up or their husbands had retired. Most of the women also made references to having had negative experiences with Germans concerning their race, nationality, or gender. Two thirds considered Germany very hostile to foreigners and half indicated that the biggest problem in the relationship between Thais and Germans is the negative attitudes of Germans towards foreigners. Yet they also expressed positive evaluations of German life:

> I like the order and the system of Germany; the laws are good here; security, quality of life, society, and the future of people, all is very good here. However, I don't like the locals in the countryside who love to put down foreigners; they are stupid. What is typical for Germans is 'Deutsch ist Deutsch'.

> The good thing about Germany is the cleanness of the cities and the discipline of the German people, but their faces are without smiles and they look coldly on children and Asian people. I miss Thai food and the Thai way of life, Thai hospitality and people helping each other. I don't miss the bureaucracy in Thailand or the transportation system. Thai people smile easily to everyone, they love to live, enjoy eating and enjoy their lives. I miss my home and I day-dream about Thailand a lot.

The women also maintain regular mail contact with their families in Thailand and manage to visit every year or every other year.

How can the women's sense of identity be characterized at this point? Emotionally they are more attached to Thailand, but pragmatically, and because of commitments at the personal level to marriage and to their German-born children, they see their immediate futures as lying in Germany. They make efforts towards being functional in German society – by learning the language, studying and, in a number of cases, seeking employment, yet they also keep most of their activities fairly closely tied to the home, are sensitive to being outside German culture and recognize German hostilities towards foreigners. To what extent their choices reflect their class position, which might best be described as from lower middle to middle middle, both in their origins and in their marital partnerships, is difficult to assess but worthy of consideration.

The concept of 'code switching', which would acknowledge competence in daily life in both cultures, together with behaviour and choices directed towards personal and familial security, but not a significant change of national identity, seems to be an appropriate representation of their preferred position.

CONCLUDING REMARKS

My case study suggests that multiple models of ideal identities for the 'other' are being advanced and developed in Germany. Official positions of 'integration' shape the legal context but do not generally resonate with the Thai women who have married German men. Yet the political reality is that the government's visions will affect the women's options in daily life and will be especially oppressive if their marriages encounter difficulties. Community non-governmental agencies are diverse in their positions but tend to see the women as victims or in need of assistance. Given the initiative of the women in contracting marriages with foreign men, their capacities seem to be underestimated, a view supported by their resistance to being patronized. For the women themselves, the preference appears to be to maintain a multifaceted and flexible position – with pragmatic valuing of security in their marriages and loyalty to their husbands and families, but sustaining an emotional identity in Thai culture.

Mirjana Morokvasic, a leading researcher on immigrant women in Europe, has recently noted that bi-national relationships and families are becoming one of the signs of transnationalization in Europe which challenge the exclusiveness of national identities and the exclusiveness of citizen rights for 'nationals'. She calls for more research on the processes by which national identities are consistently re-negotiated or camouflaged (Morokvasic 1993) . In the transnational context of the EU, with its priorities for free movement across the borders of member states and its struggles over dealing with the 'other' within as well as the 'other' without, studies that explore relationships such as those the Thai women have in Germany will become increasingly important to undertake.

NOTES

1 A total of 280 questionnaires were distributed through a variety of channels including Thai grocery stores, Thai clubs, student organizations, restaurants, German/Thai clubs, and private contacts. A missionary who had contacts with about 1,200 Thai/German couples mailed 150 questionnaires to a selection of this list. A response rate of about 20 per cent was received to the survey, including twenty-three Thai women and twenty-one German men. The questionnaires were available in Thai, German and English versions. About half of the women responding to the questionnaire had most commonly lived in Germany from three to five years, with the remainder

almost equally divided between those who had lived there more than six years and those who had been two years or less in Germany.

2 The same rules generally apply if a German woman marries a Thai man; whether a judge would grant custody to a German mother over a Thai father is uncertain, but it seems likely that the German parent would be given custody on the grounds that the child would have greater economic benefits in Germany.

REFERENCES

Abadan-Unat, N. (1977) 'Implications of migration on emancipation and pseudoemancipation of Turkish women', *International Migration Review* 2, 1: 31–57.

AGISRA (ed.) (1990) *Frauenhandel und Prostitutionstourismus: Eine Bestandsaufnahme*, Munich: Trickster.

Aziz, N. (ed.) (1992) *Fremd in einem kalten Land: Ausländer in Deutschland*, Freiburg: Herder/Spektrum Verlag.

Beauftragte der Bundesregierung für die Integration der ausländischen Arbeit-nehmer und ihrer Familienangehörigen (ed.) (1989) *Hearing zur Situation ausländischer Frauen und Mädchen aus den Anwerbestaaten*, Dokumentation Teil 1 and 2, Bonn: Bundesministerium für Jugend, Familie, Frauen und Gesund-heit.

Cohen, E. (1982) 'Thai girls and *farang* men: the edge of ambiguity', *Annals of Tourism Research* 9: 403–28.

—— (1986) 'Lovelorn *farangs*: the correspondence between foreign men and Thai girls', *Anthropological Quarterly* 59: 115–27.

Davis, F. J. and Heyl, B. S. (1986) 'Turkish women and guestworker migration to West Germany', in R. J. Simon and C. B. Brettell (eds) *International Migration: The Female Experience*, Totowa, NJ: Rowman & Allenheld.

Die Grünen (ed) (1990) *Argumente: Die multikulturelle Gesellschaft*, Bonn: Selbstverlag.

Heisler, M. O. and Layton-Henry, Z. (1993) 'Migration and the links between social and societal security', in O. Waever, B. Buzan, M. Kelstrup and P. Lemaitre (eds) *Identity, Migration and the New Security Agenda in Europe*, London: Pinter.

Hillman, F. (forthcoming) 'Immigrants in Milan: gender and household strate-gies', in C. C. Roseman, G. Thieme and H. D. Laux (eds) *EthniCity: Ethnic Change in Modern Cities*.

Humbeck, E. (1994) 'Marriage migration and changes in cultural identity: Thai women in Germany', unpublished PhD dissertation, Arizona State University.

IAF (Verband bi-nationaler Familien und Partnerschaften) (1991) *Mein Partner oder meine Partnerin kommt aus einem anderen Land*, 2nd edn, Frankfurt: Caro Druck.

Institut für Interdisziplinare Sexualforschung (1988) 'Prostitution in Thailand: Einstellung und Verhalten deutschen Männer', unpublished survey summary, Hamburg: IFIS.

Katholische Frauengemeinschaft Deutschlands (KFD) und Evangelische Frauenhilfe in Deutschland (EFHiD) (eds) (1981) *Sextourismus, Exotischer Heir-atsmarkt, Prostitution*, Dusseldorf: Hausdruck.

Kofman, E. and Sales, R. (1992) 'Towards Fortress Europe?', *Women's Studies International Forum* 15: 23–39.

Latza, B. (1987) *Sextourismus in Sudostasien*, Frankfurt am Main: Fischer Taschen-buch Verlag.

Lipka, S. (1987) *Das Kaufliche Gluck in Sudostasien: Heiratshandel und Sextourismus*, Münster: Verlag Westfalisches Dampfboot.

Morokvasic, M. (1984) (ed.) *Women and International Migration*, special issue of *International Migration Review* 18, 4.

—— (1991a) 'Fortress Europe and migrant women', *Feminist Review* 39: 69–84.

—— (1991b) 'Roads to independence: self-employed immigrants and minority women in five European states', *International Migration* 29, 3: 407–20.

—— (1993) '"In and out" of the labour market: immigrant and minority women in Europe', *New Community* 19: 459–83.

Munscher, A. (1984) 'The working routine of Turkish women in the Federal Republic of Germany: results of a pilot survey', *International Migration Review* 18: 1230–46.

Phizacklea, A. (ed.) (1983) *One Way Ticket: Migration and Female Labour*, London: Routledge & Kegan Paul.

Phongpaichit, P. (1982) *From Peasant Girls to Bangkok Masseuses*, Women, Work and Development Series no. 2, Geneva: International Labor Organization.

Ruenkaew, P. (1990) 'Leben in zwei Welten: Thailändische Frauen in der BRD', in *Liebes- und Lebensverhältnisse: Sexualität in der feministischen Diskussion*, ed. Interdisziplinäre Forschungsgruppe Frauenforschung (IFF), Frankfurt am Main: Campus Verlag.

Simon, R. J. and Brettell, C. B. (1986) *International Migration: The Female Experience*, Totowa, NJ: Rowman & Allenheld.

Statistisches Bundesamt (1989) *Statistik Zahlen für alle*, Wiesbaden: Statistisches Bundesamt.

Tichy, R. (1993) *Ausländer rein!: Deutsche und Ausländer – verschiedene Herkunft gemeinsame Zukunft*, Munich: Serie Piper.

Tubinger Projectgruppe Frauenhandel (TPF) (1989) *Frauenhandel in Deutschland*, Bonn: Verlag J. H. W. Dietz Nachf.

10

FROM INFORMAL FLEXIBILITY TO THE NEW ORGANIZATION OF TIME

Paola Vinay

As the previous chapters demonstrate, the situation of women differs greatly among the member states of the European Union. There are, indeed, noticeable differences in women's access to the formal labour market and to high level and non-traditional feminine occupations, as well as in the incidence of part-time work and of other forms of 'atypical' employment. Strong and persistent cultural differences inhibit equal opportunities for women in access to the labour market and to higher-level positions; these cultural differences also affect the division of labour and of care work within the family and hence condition women's access to the formal labour market. Moreover, deep differences can be found in the provision of public social services and in the timing and effectiveness of state legislation and policies relating to the female question, in particular to equal opportunities for women in the labour market.

Within the European Union equal opportunities for women still vary considerably among the southern and the northern countries. As Dina Vaiou makes clear (see Chapter 3), for instance, in the four countries of southern Europe informal activities are widespread, while public services that would facilitate women's access to the formal labour market are definitely inadequate. Over the last two decades, however, cultural changes have been taking place among women in these societies. Indeed, sociodemographic evidence suggests that women's attitudes towards work are changing rapidly in all four countries, even though cultural change in the society as a whole seems to come much more slowly. It therefore seems particularly relevant to analyse the problem of equal opportunities in the labour market in these countries.

This chapter will offer a detailed case study of women's employment in Italy over the last two decades in the context of related sociodemographic and cultural changes. Many of the data presented here refer to the country as a whole; however, particular emphasis will be placed on Marche, a region of central Italy that is typical of the so-called 'Third Italy'. Among the main features differentiating the social and economic structures of this region (as well as of the other regions of Third Italy, as explained below) from those of the advanced capitalist countries are the persistence and the vitality of a large sector of small industrial and craft-based firms, many of which are family concerns that perform an active function in the creation of national wealth and hence are not definable simply as marginal, residual, or pre-modern enterprises.

The region's manufacturing industry is highly specialized in the production of consumer goods such as shoes, clothing, textiles, wooden furniture and musical instruments, which have highly variable demand and involve high labour intensity in their production. The presence of a vast sector of small firms enables the regional economic system to maintain significant margins of flexibility. The dispersion of production into a multiplicity of small- and medium-sized firms and the interplay between industry and agriculture seem to be two historical constants of the Marche industrialization since its beginning in 1954–5. Although this pattern has become somewhat attenuated in the subsequent decades and updated in the course of recent modernization, it nevertheless remains a typical feature of the region's development.

THE 1950s AND 1960s: WOMEN AND THE OLD LABOUR 'FLEXIBILITY'

As is well known, industrial development in Italy has been and remains very diversified. It is widely accepted, indeed, that there are 'three Italies' (Bagnasco 1977) characterized by quite different socioeconomic formations. The regions of the north-west (the so-called industrial triangle), whose industrialization started at the turn of the century, are character- ized by the widest density of large- and medium-sized private and public firms. The regions of the south constitute the more underdeveloped part of the country incorporating extensive agricultural areas together with areas of great urbanization that are characterized by high levels of unemployment and of irregular work. Finally, the regions of central and north-eastern Italy, the so-called Third Italy, are characterized by a wide diffusion of small- and medium-sized industries.

I shall focus on this third area because the industrial model that developed there, and which has remained relatively stable, has stimu- lated the interest of a number of social scientists since the mid-1970s to the point that some Italian and foreign economists have proposed it as a model for developing countries (Fuà 1983) or as a real alternative to mass production (Piore and Sabel 1984). Moreover, in many economic and social features, the regions of the Third Italy show values that approx- imate the average national values, balancing the north and the south.

From the 1950s to the early 1980s, the Third Italy, which had been mainly agricultural, demonstrated an accelerated rate of industrializa- tion, mostly in the form of the growth of small firms: a type of indus- trialization called *dispersed* or *diffused* economy. In these regions about 80 per cent of industrial employees are concentrated in firms of fewer than

250 employees and the average industrial unit employs fewer than ten workers. This type of industrialization has been fostered by a number of economic and social conditions, including the progressive specialization in consumer goods with highly variable demand for which the more flexible small firm seems particularly functional; the artisan and mercantile tradition which has provided the local economy with a wide professional know-how; the welfare and fiscal benefits to which artisans and small farmers have been entitled; the strong social integration of the local community; and finally, the rural context and the strong ties of the local population with the land.

What it is important to stress in this chapter, however, are the implications of this type of development for women's work and contemporary changes. Among the more relevant social conditions for our analysis are the persistence of the extended family and the strong hierarchy of roles within the family. In the 1960s and 1970s, this kind of family allowed its members to undertake an heterogeneous cluster of activities for raising money and producing goods and services which would otherwise be acquired in the market. The hierarchy of roles within it allowed males in the central age brackets to be employed in the formal economy and required that women, children, young and old people work in the informal economy – both for the market and for family consumption and reproduction (Paci 1980; Vinay 1985). In other words, this type of family has favoured the labour flexibility necessary for the good functioning of the dispersed economy and of the small firm.

Research in 1975–6 in Marche, the region that we take as an example of the Third Italy, examining a sample of 650 families, highlighted the importance of the various forms of utilization of the family labour force in the dispersed economy: in informal activities for the market (irregular work, occasional work, piece-work at home, moonlighting, etc.), in household activities for family consumption (kitchen-gardening, poultry-breeding, manufacture of agricultural products and of clothes, etc.); and in the production of services for the family (housekeeping work, care work, etc.). In particular, this research made clear that the involvement of family members in the various informal activities differed according to the family life-cycle and that the presence of pre-school-age children in the family had an important influence on the organization of the family labour force, causing the exit of the wife-mother from the formal labour market and her entry into the informal one. Moreover, both the informal activities for the market and the household activities for family consumption were undertaken mainly by women. 'Black' work, for example, engaged less than 8 per cent of male bread-winners, but over 40 per cent of wives and of daughters.

Women, therefore, proved to be the chief support of the local productive and reproductive system. In fact, they accepted moving in and out of the formal labour market according to the needs for flexibility imposed by the local economy and by the family life-cycle, constituting an important labour reserve for black piece-work at home. In other words, the research suggested first that the heterogeneous cluster of activities in which families were involved guaranteed the labour flexibility so important for the local economy, and second that the forms of utilization of the family labour force in this informal economy mainly reflected the gender division of work roles (Vinay 1985). Female participation in the labour market, therefore, was deeply conditioned both by the needs of the family and by the particular form of industrial development.

THE 1970s AND 1980s: DEMOGRAPHIC AND ECONOMIC CHANGES

In the last fifteen years, however, in this area of Italy as well as in the country as a whole, important changes have occurred in the structure of the family and in the attitude of women towards paid work. The historical analysis of data shows a trend towards smaller families, while divorce rates and the incidence of children born outside marriage and of civil marriages are increasing (Table 10.1). Moreover, there has been a sharp reduction in marriage rates, in birth rates and in fertility and total fertility rates. The total fertility rate in Italy, for instance, for women aged 15–49 decreased from a mean number of 2.4 children per woman in 1961 to 1.3 in 1987; in the Marche it decreased from 2.0 to 1.2. Such rates are among the lowest in Europe today. These data suggest that contemporary women, while not renouncing maternity, choose to control the time of reproduction and the number of children.

Another relevant change is the increasing access of women to higher education, which has totally neutralized the previous gender differences in this field. If we consider the age group 25–9 years, that is, relatively young people who should have finished their formal education, we find that the percentage of females with a secondary education and also with a university degree is higher than that of males, particularly in Marche (Table 10.2). This means not only that among the younger generations the historical disadvantage for women has vanished, but that a greater proportion of females today reach higher education than do males. A similar process has been identified among Portuguese women of the same age bracket (see Chapter 7). Education, indeed, is seen by women as the main route to emancipation and to finding a good job. It is important to suggest, however, that this longer schooling of women can also be related to the fact that young males have more chances of finding a job or inheriting the family firm than young females.

Women's equal access to higher education, together with the reduction of the fertility-natality rates, contributes to the emergence of their changing attitudes towards work. Indeed, the data on the female component of the labour market show a sharp increase in female activity and employment rates. However, female unemployment in Italy

Table 10.1 Demographic indices for Italy and Marche, 1961–87

	1961	*1971*	*1981*	*1987*
Italy				
Divorce rate (1)	–	3.2	2.0	4.7
Civil marriages (2)	1.6	3.9	12.9	14.7
Children born outside marriage (3)	24.0	24.0	43.0	58.0
Marriage rate (4)	7.9	7.5	5.5	5.3
Fertility rate (5)	73.0	68.7	–	38.9
Total fertility rate (6)	2.4	2.4	1.6	1.3
Birth rate (4)	18.4	16.8	11.0	9.6
Infant mortality (3)	40.1	28.3	14.1	10.1
Abortion (3)	–	–	360.8	374.2
Voluntary abortion (7)	–	–	–	304.4
Marche				
Divorce rate (1)	–	1.4	1.2	1.9
Civil marriages (2)	0.7	1.9	7.9	9.3
Children born outside marriage (3)	9.0	120.0	20.0	32.0
Marriage rate (4)	8.5	6.9	5.2	4.7
Fertility rate (5)	60.2	57.9	41.7	34.7
Total fertility rate (6)	2.0	2.1	1.4	1.2
Birth rate (4)	15.8	14.3	9.8	8.2
Infant mortality (3)	28.0	16.8	11.2	9.9
Abortion (3)	–	–	391.7	359.0
Voluntary abortion (7)	–	–	–	275.6

Sources Istituto Centrale di Statistica (ISTAT), *Censimento Generale della Popolazione*, 1961, 1971, 1981; ISTAT, *Sintesi della Vita Sociale Italiana*, 1990; ISTAT, *Annuario Statistico Demografico*, 1988; ISTAT, *Annuario Statistico Italiano*, 1990; ISTAT, *Statistiche Sociale*, 1981; ISTAT, *Popolazione e Movimento Anagrafico dei Communi*, 1990; data processing by Paola Vinay

Key
(1) per 10,000 inhabitants
(2) per 100 marriages
(3) per 1,000 born
(4) per 1,000 inhabitants
(5) live births per 1,000 women 15–49 years of age (for 1961 and 1971 the data refer to the periods 1960–2, 1970–2; for 1987 the data refer to 1987–8 for Italy and to 1989 for Marche)

(6) mean number of children per woman 15–49 years of age
(7) per 1,000 live births (the data for voluntary abortions refer to 1989)

(also in the Third Italy) is still very high. As a matter of fact, in Italy as in the other southern European countries, the female unemployment rate is more than double the male one and in Marche is nearly three times as high (Table 10.3).

Nevertheless, in spite of the high level of unemployment, women remain in the labour market; and this is true not only for the younger women but also for those in the central age brackets and with school-age

Table 10.2 Educational level by sex and age groups for Italy and Marche, 1989
(percentages)

Italy	14–64		>64		25–9		Total	
	M	F	M	F	M	F	M	F
University	5.1	3.7	4.0	1.0	5.2	5.2	4.9	3.2
Secondary	21.8	20.9	6.6	4.3	36.6	38.2	19.4	17.7
Primary	41.0	36.0	11.0	7.9	48.3	45.3	36.3	30.4
Interrupted primary or illiterate	32.1	39.4	78.4	86.8	9.9	11.4	39.4	48.7
Total	100	100	100	100	100	100	100	100

Marche	14–64		>64		25–9		Total	
	M	F	M	F	M	F	M	F
University	4.8	3.7	2.6	0.6	5.6	5.9	4.4	3.0
Secondary	20.7	21.7	5.5	3.6	36.1	43.5	17.9	17.8
Primary	37.6	31.1	6.0	4.2	52.5	43.9	31.7	25.1
Interrupted primary or illiterate	36.9	43.5	85.9	91.6	5.8	6.7	46.0	54.1
Total	100	100	100	100	100	100	100	100

Source Istituto Centrale di Statistica (ISTAT), *Rilevazione Trimestrale delle Forze di Lavoro*, Media, 1989

children. Indeed, research in three cities of our region on women with children in primary school showed activity rates between 75 and 80 per cent (Vicarelli 1991). If we analyse the specific activity and unemployment rates by age brackets (Table 10.4), we can see that in 1989 the female unemployment rate was always at least double the male, and even persists at a relatively high level in the middle age brackets, that is, when reproductive activities are particularly demanding. (Moreover, although there is a clear divergence between male and female activity rates starting from the 25–9 age bracket, the female activity rate today remains over 70 per cent until 40 years of age and over 60 per cent between 40 and 50 years.) Figure 10.1 shows the difference between male and female activity rates according to specific age brackets for 1975 and 1989. It demonstrates that in central Italy, as well as in the country as a whole (see Abburrà 1989), the profile of women's participation in the labour market, though clearly lower than men's, is gradually taking the shape of the typical male profile, whereas until 1975 it had a sharp fall after the age bracket of 20–9 years.

Table 10.3 Activity and unemployment rates by sex for total population of Italy and Marche, 1975, 1989, 1990, 1991: percentage distributions of the labour force and specific indicators for married women, Italy and Marche, 1989

| | Activity rate | | | | Unemployment rate | | | |
	1975	1989	1990	1991*	1975	1989	1990	1991*
Italy								
Total	35.7	42.0	42.0	42.5	3.3	12.0	11.0	10.9
M	52.7	54.5	54.4	54.9	2.8	8.1	7.3	7.5
F	19.7	30.2	30.3	30.7	4.6	18.7	17.1	16.8
Marche								
Total	40.9	45.5	44.8	44.8	2.3	7.5	6.6	6.7
M	55.2	54.5	54.9	55.8	1.6	4.4	3.7	4.1
F	27.4	36.0	35.4	34.6	3.6	11.9	10.9	11.1

1989: Distribution of the labour force	*Italy*	*Marche*
Total labour force	100.0	100.0
% Male labour force	63.1	59.5
% Female labour force	36.9	40.5
% Married female labour force	22.1	26.9
Specific activity rate married women	35.8	**
Specific unemployment rate married women	12.4	8.1

Source Istituto Centrale di Statistica (ISTAT), *Rilevazione delle Forze di Lavoro*, Media, 1975, 1989, 1990, 1991; our data processing
Key * For 1991 the source is 'Foglio di informazioni ISTAT', 11–02–92.
** *At the regional level the specific activity rate for married women is not available.*

Table 10.4 Specific activity, employment and unemployment rates by sex and age brackets, Marche, 1989

| Age groups | Activity rate | | Employment rate | | Unemployment rate | |
	M	F	M	F	M	F
14–19	27.0	25.7	23.0	18.0	15.1	30.0
20–4	71.7	71.0	60.4	53.8	15.8	24.2
25–9	90.7	77.1	80.1	62.0	7.3	19.6
20–9	81.3	74.1	72.3	57.9	11.0	21.8
30–9	97.6	73.6	94.8	66.9	2.9	9.2
40–9	95.9	60.2	94.2	56.8	1.8	5.5
50–9	82.9	37.6	81.3	36.7	1.9	2.5
14–59	81.2	57.0	77.3	50.0	4.8	12.4
60–4	44.0	15.2	43.6	15.0	1.0	1.4
14–64	77.6	52.8	74.0	46.5	4.5	12.0
65 and over	11.0	3.3	10.9	3.3	–	–
Total > 14	65.2	41.7	62.3	36.7	4.4	11.9

Source Istituto Centrale di Statistica (ISTAT), *Rilevazione delle Forze di Lavoro*, Media, 1989; our data processing

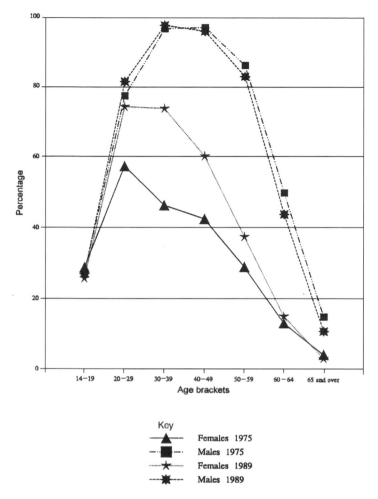

Figure 10.1 Specific activity rates by age and sex, Marche, 1975 and 1989
Source Istituto Centrale di Statistica (ISTAT), *Rilevazione Trimestrale delle Forze di Lavoro,*
Media, 1975, 1989

THE 1990s: A NEW CULTURAL MODEL
(VERSUS AN OLD FAMILY POLICY)

Contrary to what happened in the past, women today remain in the labour market in spite of the high rates of unemployment; that is, they are not discouraged by the selectivity of the demand favouring men and are not willing to give up their professional career and their economic independence. A new cultural model is emerging in the regions with a dispersed economy as well as in the country as a whole with respect to women's attitudes towards family and work. In the 1960s and 1970s, as we have seen, women were deeply conditioned by the needs of the local labour market and of the family life-cycle. Their cultural attitudes then placed a higher priority on the needs of the family than on paid employment. The prevailing cultural model for women's work was one of family priority.

Now, on the contrary, among the younger generations a new cultural model is clearly emerging: women are no longer willing to give up their professional career and their economic autonomy. This pattern does not imply that they place a higher priority on work than family, however. Indeed, the originality of their new attitudes lies precisely in the fact that they do not want to choose between work and family. Rather they will not renounce either one, but want society to acknowledge the equal dignity of work for the market and caring work for the family. The prevailing cultural model for women's work today is therefore a plural model according to which self-fulfilment results not only from professional activities but also from commitment in the other important dimensions of life. This cultural model differs considerably from the typical male mono-cultural model according to which self-fulfilment results mainly from the professional career (Manghi 1987).

This non-choice between family and work by the modern woman and, therefore, her double day in work for the family and for the market means a limited participation in public life and, for many women, an almost complete lack of free time. The new cultural model of women, indeed, has not been accompanied by a change in society as a whole with respect to equal opportunities in the access to labour market, to free time and to public life. In fact, in spite of the increase in employment rates, unpaid housework and care work is still mainly done by women. Recently research in Marche, for instance, has shown that women with school-age children, no matter what their professional role, devote three to four hours per day to their children and four to five hours per day to housework; the research has shown, moreover, that the redistribution among sexes of this unpaid work is far from being achieved (Vicarelli 1991).

Italian data on gender differences in the use of daily time show that men's patterns do not vary significantly according to the different types of families in which they live, while women's are significantly affected by the presence of children in the family and of a husband. Indeed, male participation in housework and childcare is so marginal that the lack of a husband in the family reduces women's overall housework. Since women have to take care of their husbands also, it seems that (as it has been remarked) in Italy 'men are advantaged by the presence in the family of a partner woman, while women are advantaged by the absence of a partner man' (Sabbadini and Palomba 1993: 224 and Table 6.6). Further, the gender segregation of house and care work persists when women are employed and increases with the number of children (Sabbadini and Palomba 1993: Tables 6.7 and 6.9).

These changes have occurred in a context in which the Italian state has not created policies to support women's dual roles. Indeed, in spite of the increased access of women to the labour market and of the growing demand for caring services consequent to greater incidence of aged and non-self-sufficient people, the Italian state very often chooses to provide financial assistance to families rather than social services. Not only is social expenditure per capita relatively low, but within it the share of monetary transfers to families in 1988 was about 43 per cent (Paci 1991).

Further, in recent years in order to cut welfare expenditures the government has reduced the national funds devolved to municipalities for social and welfare services. Because of the lack of a clear and definite national policy, the geographical distribution and the quality of social services in Italy differ greatly from region to region and, even within the same region, from one municipality to another. The distribution of social services and provision of infrastructure for children, elderly and non-self-sufficient people is therefore very uneven depending on the budget of the municipality and on the political type of local government. Social and health services as well as schools are concentrated in the larger municipalities of the northern and central regions of Italy, while they are almost absent from the smaller ones, particularly in the inner mountain areas. Two examples from Marche illustrate this situation. In 1989, only one fifth of the municipalities in the region had a nursery, while only 3 per cent of those with fewer than 5,000 inhabitants had one. The overall coverage for children of nursery age (under 3 years) was less than 9 per cent and for every 100 registered children, 46 were on the waiting list to enter nurseries (David 1991). Research on handicapped people in four municipalities of the region showed that the care work needed by these disadvantaged citizens is guaranteed almost exclusively by the family (that is, by the women – wife, mother, sister), public home care services being totally unavailable (Vinay 1990). It is interesting to recall here what Fagnani says about France (see Chapter 6) with reference to the relation between high fertility, full-time employment rates for women and adequate policies and programmes to aid couples to reconcile professional and familial work in that country. In Italy, by contrast, the lack of social policies and of adequate social services may account for the sharp fall in fertility rate and for the persisting inequality of work opportunities for women.

WOMEN AND WORK TODAY:
STILL AN UNFAIR DISTRIBUTION

There is no doubt that the lack of social services and policies, as well as the unequal division among the sexes of care work, has marked consequences over women's professional careers compared with men's. The need for time leads women towards the choice of occupations with work schedules compatible with the double day and compels some of them to accept atypical forms of employment such as seasonal, casual, irregular and part-time employment. Although in Italy (as in the other southern European countries) part-time contracts are less common than in the northern countries of Europe (only 3.4 per cent of all employment in 1988), this type of employment mainly involves women in the central age brackets when double presence problems are more pressing. Recent research in Marche (Bugari 1991) showed that the majority of women employed under part-time contracts were compelled into this choice by their family duties.

As far as employment in the underground economy is concerned (which was so important in the 1960s and 1970s for the diffused economy) no statistical data are available for recent years. The modification of some of the conditions which had favoured it, together with the decrease in industrial home-working and with its legal regulation, resulted in a decrease in irregular work as a whole. It is known that irregular work is still relatively frequent in personal services (paid house-keeping, baby-sitting), in retail trade, in tourism and recreation and in small industrial firms, mainly in the clothing industry. It is interesting to recall in this connection that between 15 and 18 per cent of mothers of school-age children interviewed in 1991 in three towns of Marche held a job without any contract or were not registered as self-employed (Vicarelli 1991). Women's continuous need for time, moreover, makes them unavailable for extension of the work schedule and for overtime work; rather it leads them to favour reduction of the daily work schedule and forms of individual flexibility (Bugari 1991).

In short, although the attainment of educational equality and the emergence of new professions have favoured female employment and have opened new prospects to women, the majority of them are still employed in the lower functional levels – the less professionalized and worst paid ones, and the higher one climbs the professional hierarchy the lower is the proportion of women (Materazzi 1991). In addition, in the majority of female jobs, career possibilities are quite limited. Both in the public and private sectors, it is possible to reach the higher functional levels only if one is ready to be available full-time for work. (As André underlines in Chapter 6, availability in terms of time is also an important requirement for employers in Portugal.)

The consequence of this horizontal and vertical segregation of female employment, as was shown by a recent study of unions in the country as a whole (Altieri and Schipani 1991), is a sharp difference in average earnings among the sexes in all economic branches so that in 1987 the average female earnings were only 76 per cent of the average earnings for men. For those women who are still compelled by the conditions of the labour market and by the needs of the family to accept part-time or temporary jobs, or jobs not guaranteed by collective bargaining, earnings are significantly lower and the extent of inequality between the sexes has important social consequences.

THE ISSUE OF REORGANIZATION OF TIME

The contradiction between the changed cultural model of women and the persistence of old discrimination in Italian society is clear. The attainment of educational equality and the desire of women to experience all the dimensions of life are at odds with the lack of equal opportunities in the labour market, with the lack of social services for the family and with the gender division of work roles. It is therefore essential today to involve the whole society in the cultural change started by women. From this point of view the debate recently fuelled in Italy on the reorganization of time schedules for work and daily life is highly relevant. This debate is centred, more generally, on the need for more flexible work schedules in contemporary post-industrial society and on the acknowledgement of the social value of caring work within the family and its equal division among the sexes if a full and equal participation in all dimensions of life is to be enjoyed by all. More specifically, this debate focuses on the revision of social services policy and on the schedules of all urban public and private services so that they be compatible with work schedules.

Although the debate on the reorganization of time is fuelled by the growing unease that women experience, it is important to recall that there are several structural and cultural reasons which make the organization of time brought about by industrialization increasingly inadequate in Italy as well as in the other developed countries.

As is well known, industrial development – in particular Fordism – basically standardized time schedules of all citizens and of all the sectors of economy: all are at work, eat, watch TV and go to bed at the same time of the day, and all are on holiday at the same time of the year.

> In other words, industrialization and urbanization created a dominant, linear, rational time . . . accompanied by the discontinuous, dispersed, apparently still natural time of family life and of women. Indeed, researches have shown that the normative calendars, the regular, linear scanning of time introduced by the new industrial societies have been and still are based on the disorder of women's time . . . and on their ability to combine together activities and needs.
>
> (Vicarelli 1992: 6 and the bibliography there cited)

In today's post-industrial society, when the service sector prevails and when rigid programming can be overcome by new flexible technologies, all the contradictions linked to the rigid and standard organization of time emerge. Since the 1980s de-synchronization of social activities is taking place; with the increase in female participation in the labour market the reorganization of time has become an important issue if equal work opportunities among the sexes are to be attained. Indeed, inequality in the use of time is now perceived with an increasing sense of unease and of injustice, particularly by women. Research has shown that in some working contexts anxiety because of lack of time is two and a half times more frequent in women than in men. This fact explains why the request for the reorganization of de-synchronized time schedules for work and daily life primarily originates from women (Chiesi 1992 and the bibliography there cited).

It is important to recognize that the reorganization of time schedules must be undertaken systematically because of the interdependency of the various schedules with one another. For example, as has been suggested (Manacorda 1991), a better distribution of the time of entrance to school could contribute to reducing traffic peaks and pollution, but the school schedule is linked to the work schedule of parents and to the schedules of buses and trains; moreover, the need for flexible work schedules contrasts with the need for some services to be available to the public at specific times. There is, therefore, a chain of interdependency which must be carefully considered in order to avoid negative consequences. For instance, we should not overlook that the majority of workers in public services are women whose interest as workers in the present Italian 8.00 a.m.–2.00 p.m. work schedule could be at odds with their interest as consumers in having wider access to services at different times of day and week.

Since the second half of the 1980s the reduced flexibility of the female labour force and the difficulty of combining the double role of worker and user of services have fuelled several experiments in different Italian cities aiming to improve the connections between urban time schedules and the needs of citizens. The most significant example is the experimental project started in Modena, a town of central Italy whose mayor was a woman. This project has shown that it is possible to guarantee a better interaction between the schedules of all public and private services, to extend and qualify, at reasonable costs, local public services so as to make them consistent with the new social demand and to speed up bureaucracy (Rinaldi 1991). This first project has been followed by others in various Italian cities (Milan, Genoa, Perugia, Cagliari, Livorno, Ravenna, Vicenza, Orvieto and others) favoured also by a new law on the organization of municipalities (Law 142, 1990) which entrusts to the mayor the duty of co-ordinating the schedules of public and private services.

CONCLUDING REMARKS

To sum up, in the last fifteen years, as we have seen, women's conditions in the regions of the diffused economy as well as in the country as a whole have changed. Although women have access to jobs formerly considered male, and to some emerging professions, the majority of work opportunities open to women in the service sector are still too often characterized by low functional levels and by low wages. This definitely contrasts with the high levels of education women have acquired. Nevertheless, Italian women seem no longer to be willing to give up their professional career and economic independence, even though their desire to enter the labour market does not always find adequate answers.

In the past the dispersed economy favoured the work of women, albeit marginal and precarious work. Still linked to a model of family priority, they accepted the casual, irregular jobs necessary for the flexibility of the economy. Today this type of flexibility is less important. Women have acquired equal opportunities in education and access to the formal labour market. They have developed a plural model which gives value both to family and work.

In these new conditions flexibility must be organized. It cannot be as informal as it was in the 1960s and 1970s. This is why it has become so important in Italy today, both for the women's movement, and also for many unions and political parties, to intervene consciously in the reorganization of the time of work and in the schedules of all urban public and private services. (These issues have acquired recently new relevance for the social partners in the face of growing unemployment. In Italy, the reduction of the time of work for all has become a matter of collective bargaining in the major industrial firms like FIAT and Olivetti.) The debate is just beginning, but it must be pursued rapidly in order to meet the needs of organizing everyday life and to guarantee equal opportunities among the sexes with respect both to work opportunities and to participation in all the other dimensions of life, family, culture, political and social commitment and time *per se* whose unequal distribution is synonymous with injustice, and, in the end, interferes also with complete self-fulfilment of both sexes.

REFERENCES

Abburrà, L. (1989) *L'occupazione Femminile dal Declino alla Crescita*, Turin: Rosemberg-Sellier.

Altieri, G. and Schipani, S. (1991) 'Le differenze retributive tra uomini e donne', in S. Patriarca (ed.) *Redditi Retribuzioni e Ineguaglianze: Il Lavaro Dipendente negli Anni Novanta*, Rome: EDIESSE.

Bagnasco, A. (1977) *Le Tre Italie*, Bologna: Il Mulino.

Bugari, I. (1991) 'Le donne e il tempo di lavoro: un indagine in Provincia de Pesaro', *Prisma* 23: 59–69.

Chiesi, A. M. (1992) 'Politiche del tempo, metodi di ricerca e strumenti di intervento', Milan: *Ires Papers* 19: 1–27.

David, P. (1991) 'Le Politiche Sociali', *Osservatorio Regionale sul Mercato del Lavoro*, special issue of *Donne al Lavoro* 18: 183–99, Ancona: Regione Marche.

Fuà, G. (1983) 'L'Industrializzazione nel Nord-Est e nel Centro', in G. Fuà and C. Zacchia (eds) *Industrializzazione senza Fratture*, Bologna: Il Mulino.

Manacorda, P. (1991) 'Piano regolatore degli orari', in L. Balbo (ed.) *Tempi di Vita*, Milan: Feltrinelli.

Manghi, S. (1987) *Il Barone e l'Appredista: Ricerche sulla Condizione Accademia nell'Università di massa*, Milan: Franco Angeli.

Materazzi, M. R. (1991) 'Donne e mercato del lavoro nelle Marche: Alcune Riflessioni', *Osservatorio Regionale sul Mercato del Lavoro*, special issue of *Donne al Lavaro* 18: 31–60, Ancona: Regione Marche.

Paci, M. (ed.) (1980) *Famiglia e Mercato del Lavoro in una Economia Periferica*, Milan: Franco Angeli.

—— (1991) 'Classi sociali e società post-industriale in Italia', *Stato e Mercato* 32: 199–217.

Piore, M. J. and Sabel, C. F. (1984) *The Second Industrial Divide: Possibilities for Prosperity*, New York: Basic Books.

Rinaldi, A. (1991) 'Tempo e orari nella città: l'esperienza di Modena', *Prisma* 23: 77–82.

Sabbadini, L. L and Palomba, R. (1993) 'Differenze di genere e uso del tempo nella vita quotidiana', in M. Paci (ed.) *Demensioni della Desuguagliaza Sociale*, Bologna: Il Mulino.

Vicarelli, G. (1991) 'Strategie e tempi della vita quotidiana: una ricerca nelle Marche', *Prisma* 23: 43–58.

—— (1992) 'Governare il tempo: la regolzione degli orari come nuova politica di governo della città', unpublished research project, ANCI Marche, Università di Ancona.

Vinay, P. (1985) 'Family life cycle and informal economy in Central Italy', *International Journal of Urban and Regional Research* 9, 1: 82–98.

—— (1990) 'I Portatori di handicap nell'Associazione dei Comuni Ambito Territoriale No. 16 delle Marche: quantificazione e attitudini lavorative', unpublished research report, Ancona: Prospecta.

—— (1992) 'Donne e lavoro: dalla flessibilità informale alla riorganizzazione dei tempi', *Prisma* 25: 68–77.

11

CITY AND SUBURB: CONTEXTS FOR DUTCH WOMEN'S WORK AND DAILY LIVES

Joos Droogleever Fortuijn

WOMEN'S LIVES IN THE NETHERLANDS

Dutch society presents a paradox. While the Netherlands is famous for its strong feminist movement and for tolerance about certain gender issues such as homosexuality, abortion, birth control and single parenthood, in terms of a basic 'gender contract' Dutch society maintains a 'housewife' contract. Regulations in taxation, social security and housing favour traditional one-earner families (Pott-Buter 1993). Though in several cases anti-discrimination legislation in the Netherlands is enforced by the European Union (EU) – for example, with respect to gender equity in old age and widow(er)s' pensions – in numerous areas of daily life, the housewife contract prevails. Notable among these are the legal restrictions on the opening hours of shops and public services and the lack of provision of childcare facilities. Banks close at 4.00 p.m., post offices at 5.00 p.m. The Dutch parliament struggled for many years over the issue of the statutory opening hours of shops and only recently changed their time of closing from 6.00 p.m. to 6.30 p.m. Schools are open from 8.30 a.m. to 3.00 p.m. or 3.30 p.m.; primary schools are free on Wednesday afternoons. Public childcare is scarce – a place in a daycare centre is provided for less than 5 per cent of all children under 3 years of age.

In 1960, Dutch society was made up of households following the traditional family life-cycle: young couples, families with (many) children, older couples in the empty nest stage and widows and widowers. Families operated on the housewife–breadwinner model: only 7 per cent of all married women had a paid job. Gradually the household composition has become more diversified. More and more households consist of one person: young persons leaving their parental home, people living alone throughout their adult life, divorced persons of all ages and older persons, mostly women. Since 1975 two-parent families with children have formed a minority of all households and they are no longer synonymous with housewife–breadwinner families.

For a long period of time, the rate of female labour force participation in the Netherlands was extremely low compared with the other countries of the EU. In 1973 the Dutch rate was 29 per cent, the lowest in the EU. By 1983 the Dutch rate had reached 40 per cent, but only Ireland and Spain had a lower rate among the member states. The rise in the female participation rate in the period 1983–91 was higher in the Netherlands than in all other countries of the EU, however. As a consequence, in 1991 the Dutch rate (55 per cent) reached the average European level (Organization for Economic Co-operation and Development [OECD] 1993: 192).

Nevertheless, employment does not mean the same for Dutch women as it does for those in many other countries of the EU. For the majority of Dutch women, paid work means part-time work, whereas in the other countries the part-timers form a minority of all employed women (ranging from 7 per cent in Greece to 44 per cent in the United Kingdom). In the Netherlands, by contrast, 62 per cent of all working women have a part-time job (OECD 1993: 188); 49 per cent of all women with children have a paid job; however, less than 10 per cent of all mothers work full-time.

So the paradoxes of women's lives in the Netherlands are reflected in both cultural and economic behaviour at the household level and in legislation and the provision of services by the state. Nevertheless, within the Netherlands important differences exist in the patterns of women's work and daily lives between the urbanized western part of the country and the other regions, and, within the urbanized part, between inner cities and suburbs. The purpose of this chapter is to understand how the Dutch paradox plays out in different urban contexts.

FEMINIST GEOGRAPHY AND THE CITY–SUBURB DICHOTOMY

The position of women in cities and suburbs has been a major issue in the feminist geographical debate since its origins at the end of the 1970s. Studies on the phenomenon of suburbanization in the 1950s and 1960s focused on families (for instance, Rossi's *Why Families Move*, 1955) or on men (like Whyte's *The Organization Man*, 1956). Most included some criticism of the social consequences of suburbanization (Clark 1966; Riesman 1958; Whyte 1956). In general, however, the research of that period was not aware of the gendered society. At the end of the 1970s, studies on suburbanization began to pay explicit attention to the specific position of women in suburbs (Michelson 1977; Popenoe 1977; Rothblatt et al. 1979) and a few years later the first critical comments from a clearly feminist perspective appeared, contrasting suburban and city life and highlighting the ambiguities of suburban living for women.

Saegert's (1980) influential work posed city and suburb as a contradictory dichotomy, paralleling related dichotomies such as public–private, culture–nature, production–reproduction and masculine–feminine.

In societies with patriarchal gender relations, suburbs were interpreted as worlds of women, for women and against women. Others portrayed suburbs as the materialization of a gender division of labour in which men fulfil roles as breadwinners while women are the keepers of the home. Yet, at the same time, as McDowell (1983) and MacKenzie (1989) have shown clearly, the high costs of suburban living (housing, consumption, transport) are undermining this gender division, so that paid work by suburban women is both necessary and impossible. Seeing the suburban milieu as negative, Wekerle (1984) argued: 'A woman's place is in the city.' The density, the heterogeneity, the diversity of functions, the public transport and the tolerant social climate facilitate the participation of women in public life and make the combination of paid and unpaid work possible.

Recently, feminist geographers have emphasized the diversity in residential preferences of different women according to differences in life-course, class, life-style and (sub)urban background (Cook 1988; Fagnani 1993; Fava 1988; Saegert 1988). This chapter takes up the new perspective to examine the changing conjunctions between context – city and suburb – and the life-course in the household, work and residential histories of fifty dual-earner families in the metropolitan region of Amsterdam. After providing a short description of the history of suburbanization in the Netherlands, it will focus on the case of dual-earner families.

SUBURBANIZATION IN THE NETHERLANDS

Suburbanization in the Netherlands essentially started at the end of the nineteenth century with the motorization of transport. During the eighteenth century there had been some seasonal commuting: in the winter the families of rich merchants and bankers lived in town, and in the summer the wife, the children and the servants lived in the country house where the husband joined his family occasionally. Subsequently, innovations in transport – trains, street-cars and private cars – made daily commuting and suburban residence possible for the rich. Mass suburbanization, however, did not begin until the 1960s, when the postwar baby-boom generation reached adulthood and needed housing. So far the story is the same as in other European countries, but three factors distinguish the process of suburbanization in the Netherlands from that in other places:

- the structure of the urban areas;
- the state intervention;
- the participation of women in the labour market.

The Netherlands is a highly urbanized country with a high population density, lacking large metropolises, but with a ring of cities of less than 1 million inhabitants each in the western part of the country, the Randstad.

In 1966, the national government published the Second Report on Urban and Regional Planning which laid the foundation for the current suburbanization pattern in the country. 'Bundled deconcentration' formed the core concept: no suburban sprawl, the American nightmare, but a clustering of population and employment in growth centres and new towns outside the urban ring, in order to retain a 'green belt', the *Groene Hart* (green heart) inside.

Because of an absence of steering instruments, the suburbanization policy was not successful with respect to employment, which developed beyond the influence of national planning and was directed to non-planned locations. Many firms left the inner cities and moved to more spacious and accessible locations in the *Groene Hart*, which as a consequence became less 'green'. Local authorities welcomed the arrival of firms and denied national planning. By contrast, the policy succeeded with respect to the suburbanization of population. The Dutch national government has many steering instruments for housing. Hundreds of thousands of houses have been built in the growth centres, mostly social rental housing (i.e. subsidized rental housing for lower income households; in the Netherlands this represents 34 per cent of all houses). The steady rise of incomes and an extensive system of housing subsidies made suburban houses affordable for many people. Families of all social classes moved to the suburbs, except for the very poor, who stayed in bad housing conditions in the cities, and the very rich, who stayed in the cities or lived in the luxurious older suburbs. The cities were attracting fewer and fewer residents while the suburbs boomed. As a consequence, population and employment suburbanized in opposite directions resulting in long commuting distances for workers – although absolute travel distances in the Netherlands are shorter than in regions like Paris and London.

Economic, political and social-demographic reasons prompted a change in this pattern in the 1980s. Growth in incomes slowed down and more and more households became dependent on social security. Politicians who realized that space, energy, resources and time were being wasted, wanted to reinforce the position of cities. Immigrants from the Mediterranean and the West Indies came to live in the cities. Household composition diversified. More and more people chose urban residence. Thus, since 1985 the city populations have increased. There is a growing differentiation between cities and suburbs with respect to the household composition and life-styles of the inhabitants. And this differentiation has important gender implications.

SUBURBAN AND URBAN HOUSEHOLDS: SITE AND SITUATION

Two concepts – site and situation – are useful in our attempt to understand the meaning of place of residence for Dutch women's work and daily life. Site characteristics refer to the internal qualities of a residence:

the accommodation for the persons in the household, their home-based activities and their goods. Situation characteristics refer to the locational qualities: the opportunities to do activities outside the home, the availability and accessibility of employment services, of recreational opportunities and of social relations.

If we consider the residential preferences of households, urban households are more situation-oriented and suburban households are more site-oriented. One-person households and working or student couples are examples of situation-oriented, urban households, while traditional families with children are site-oriented, suburban households. Some household types, such as dual-earner families or working lone parents, are ambivalent. As a family with children they are site-oriented, as working persons they are situation-oriented. For their children they want a large and comfortable house with room to play outside. As working parents they emphasize locational aspects: the availability and accessibility of jobs and facilities, such as childcare services. Let us look in more detail at one of the ambivalent household types: dual-earner families with children.

LIFE-HISTORIES AND HOUSEHOLD ARRANGEMENTS OF DUAL-EARNER FAMILIES IN THE AMSTERDAM REGION

In 1987, I collected time-space budgets and life-histories of fifty dual-earner families with children under 15 years of age. Both partners, wife and husband, kept a diary for four days. One of the partners (half of them the male, half of them the female) was interviewed about their work, household and residential histories. The respondents were selected in three areas: an inner city area, Amsterdam-South (sixteen), a newly built residential area, Gaasperdam (sixteen) and a new town, Almere (eighteen). All areas are mixed in their socioeconomic level. The register of population of Amsterdam and Almere formed the base of the selection. Since the municipality does not register the labour participation of the population, a large sample was used for a pre-selection of the required household type.

The inner city study area, Amsterdam-South, consists of two adjacent neighbourhoods: one with small apartments and bad housing conditions with a lower socioeconomic status population, and the other a high-status residential area with large, expensive condominiums, high-quality employment and many supportive facilities, such as shops open in the evening, catering services, laundries and childcare facilities. Many singles of all ages, older couples and middle-aged dual-earner households with and without children live in Amsterdam-South. The dual-earner families in this study turned out to live in the high-status neighbourhood.

Gaasperdam is a recently built residential area in south-east Amsterdam, 10 km outside the city centre, with excellent connections and a high-quality office area nearby. It is a quiet neighbourhood with flats and single-family dwellings, partly social rental housing, partly private rental or owner-occupied houses. The population is mixed in ethnic, socio-economic and household composition.

Almere is a 'new' town: a completely new, planned settlement in the polder Southern Flevoland, about 30 km east of Amsterdam. Thirty years ago it was just water, but in 1976 the first houses and a temporary shop were built on the new land. Now it is a suburb with 78,000 inhabitants. Most houses are single-family dwellings, partly social rental housing, partly moderately priced owner-occupied. Almere has some employment opportunities, but the majority of the labour force commutes to Amsterdam or elsewhere. There are many recreational opportunities and facilities for daily use. Most households are actual, future, or former traditional families: one-earner families, young couples and divorced one-parent families. The socioeconomic status of the population is mixed and is neither very high nor very low.

Dual-earner families in Amsterdam-South

The typical dual-earner family in the inner city area of Amsterdam-South is the dual-career type: a family with two professionals (Rapoport and Rapoport 1969). Their life-histories are characteristic of people of their class. (See Figure 11.1.) Women and men leave their parental home when starting their university studies at the age of 18–19. As singles they rent a room in a students' house or a private room in the inner city. After several years they start living together with their partner in a small, uncomfortable apartment in the inner city or in a modern flat in an outer part of the city. Shortly before or after graduation they enter the labour market with temporary, part-time work. A few years later both partners get a job in a male-dominated sector; they have a full-time position with career prospects. They buy a large condominium in one of the most expensive neighbourhoods of the inner city.

When the wife is over 30, they start a family. The birth of the child is carefully planned and forms the last stage in the process of social settling. The dual-career family is born. However, it is the Dutch version of the dual-career family. Working mothers in the Netherlands seldom have a full-time job. And even full-time working mothers seldom work full-time throughout their life: during the period when they have two or more young children they work part-time. In some households, the husband has a demanding career, a more than full-time professional or managerial job, and the wife works 25–30 hours a week. In other, more symmetrical families, both partners have a full-time job, but in practice they work less than 40 hours a week.

In all households the paid work of wife and husband has priority and domestic work is arranged around it, using an extensive network of paid help: nannies, child-minders, day-care centres, domestic help and domestic services, like laundries and catering. But even in these careerist

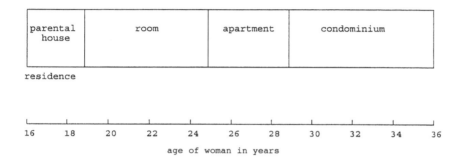

Figure 11.1 Life-path of a dual-earner family, inner city
Source Field research, Institute of Human Geography University of Amsterdam

families, both partners make small adjustments between their paid work and their domestic work. In some households the partners have partly overlapping working hours or working days; others leave home late, return early, or do part of their work at home. In most households both partners work close to home and go by bicycle. They take turns in occasional use of the family car.

These couples are highly satisfied with their house and neighbour-hood. They want to go on living in this neighbourhood for a long time. They appreciate its locational qualities, such as the availability and quality of facilities and the nearness of the city centre, as well as its good atmosphere. Negative aspects of the neighbourhood are the pollution, the traffic and the lack of safety. Nevertheless, they mention inner city areas as the ideal context for their way of life.

Dual-earner families in Gaasperdam

The dual-earner families in the inner suburb of Gaasperdam are middle-class families with a rather high educational level, but with relatively low-paying professions like primary or nursery school teaching, nursing, or community work. At the age of 20 (see Figure 11.2) the young woman or young man leaves home to live as a student in a rented room. After a few years she or he forms a couple with a partner. During the period of their education and entry into the labour market they move several times from one apartment to another. The first child is born in this period. When both partners have a permanent job, they move to a single-family dwelling in Gaasperdam; the second baby is on the way or newly born.

With the birth of the first child they try to maximize their income by working as many hours as possible, while still taking responsibility for household tasks and childcare themselves. They want to raise their children with a minimum of outside help. Unlike the dual-career families in Amsterdam-South, they lack the means for extensive paid help. In many cases the house is heavily mortgaged. Their household arrangement can be typified as a sequential scheduling strategy (Pratt and Hanson 1991; Rose and Chicoine 1991) – the parents leave home in turns to work full-time or part-time. Usually one of the parents is at home with the children. Although the women do most of the domestic work, the division of childcare is rather symmetrical.

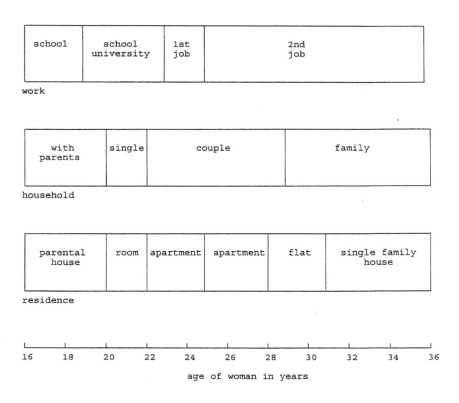

Figure 11.2 Life-path of a dual-earner family, inner suburb
Source Field research, Institute of Human Geography University of Amsterdam

These couples are less satisfied with their house and neighbourhood than the families in Amsterdam-South and Almere. They value the quiet and spacious neighbourhood with good accessibility, but they complain of the poor facilities and the monotonous and sterile architecture. Almost half of them plan to move elsewhere, in most cases to a better suburban environment.

Dual-earner families in Almere

Unlike those in both Amsterdam-South and Gaasperdam, the typical dual-earner family in the new town of Almere is a working-class or lower-middle-class family. Their life-histories reflect the characteristics of their class. (See Figure 11.3.) Young women and men leave their parental home after finishing vocational training and enter the labour market at the age of 20-1 to form a household with their partner in a modern flat in the city. In some cases, the children are born in the flat and the family moves to a single-family house in Almere when it is complete. Others move as a dual-earner couple to Almere to rent or buy a single-family house.

school	1st job	2nd job	3rd job	4th job	housewife	5th job

work

with parents	couple	family

household

parents house	flat	single family house

residence

| 16 | 18 | 20 | 22 | 24 | 26 | 28 | 30 | 32 | 34 | 36 |

age of woman in years

Figure 11.3 Life-path of a dual-earner family, new town
Source Field research, Institute of Human Geography University of Amsterdam

When the wife is in her mid-twenties, the first child is born. The husband becomes an absent father with long working hours and commuting time. The family car is his car. When he is at home, he helps with dish-washing and puts the children to bed, but in his daily life there is no interweaving of paid work and domestic work: he is a traditional breadwinner. The wife does not simply resign from her full-time job with the birth of the baby. In many cases she has already been dismissed prior to taking maternity leave. She is used to changing her job every one or two years, but because of her pregnancy, she decides to wait for some time before seeking a new job. Some of these women remain at home until the youngest child is 10–12 years old and then take a part-time job. Others wait until the youngest child is old enough to attend school and then start working during school hours. Sometimes, six months or one year after the birth of the baby, she looks for a part-time evening, night or weekend job. In any case, unpaid work, in particular childcare, has first priority. In contrast to the Gaasperdam families, there is no mutual adjustment between the partners. It is the wife who arranges her paid work around the unpaid work. She minimizes outside help. Infrequently a neighbour, a friend, or, most likely, the maternal grandmother takes care of the children. Because of her class position she has not enough money for paid outside help; the household needs both wages to pay the rent or mortgage and equipment of the suburban single-family dwelling.

These families are quite satisfied with house and neighbourhood. Although they complain of the lack of facilities and public transport, the poor accessibility and the architectural monotony, they emphasize site qualities, the availability of nature and of recreational opportunities. They have no plans to leave Almere. Suburban areas are mentioned as ideal places for living.

DISCUSSION

Dual-earner families occur everywhere in the Netherlands and are not specific to either an urban or a suburban setting. The different household arrangements of dual-earner families, however, are context-specific. Site and situation are useful concepts to understand the environmental needs of families with different household arrangements. But the simple dichotomies, urban–situation and suburb–site, need some revision.

In general, the households in the inner city study area are career-oriented and emphasize the locational aspects; however, they make no concessions in site characteristics. They live in a roomy house, often with a garden. They pay dearly for the unique combination of site and situation advantages in one of the most expensive urban neighbourhoods but, given their household arrangement, they can afford this residential situation. The context is a precondition for the household arrangement and the arrangement forms a precondition for living in this context.

The households in the inner suburban study area are more ambivalent about their environmental needs. They would not make concessions to site qualities and they cannot make concessions to situation qualities. Because of their sequential scheduling strategy, the work locations of both partners have to be nearby. They lack the means to combine both site and situation qualities. The inner suburb forms a second-best option: not too far away from employment opportunities and facilities and with fairly good site qualities.

The households in the new town study area are family-oriented in their household arrangement and in their environmental claims. Because of their low educational and income level, they cannot afford a place of residence with both site and situation qualities. They have a choice between the 'bad site–good situation' of an urban flat or the 'good site–bad situation' of a suburban single-family house. Site qualities have priority. But the costs of suburbanization are visible in their daily life. Because of the long commuting distance the husband is barely available for domestic work and childcare. The wife does all the unpaid work in a context not adapted to the combination of paid and unpaid work. She minimizes her time investment in paid work. In this sense the suburban environment reinforces the asymmetry in the gender division of labour.

At this moment in the Netherlands, labour market participation is the main issue in national political and economic debates. In the Dutch economy high productivity goes hand in hand with a low participation of men and women measured in working hours. It seems, however, that the limits of the welfare state have been reached. An increase in labour market participation has first priority on the public agenda. The labour market participation of women is growing rapidly. More and more women with children will do paid work in the future: they want to do it for emancipation reasons and they have to do it for economic reasons. Little by little, the preconditions are improving. In 1990 the national government outlined a childcare policy, for both economic and emancipation reasons. In two years public childcare facilities have been doubled; no longer restricted to the cities in the western part of the country, they are now set up in all municipalities. The rise in dual-earner families is occurring in all parts of the country. For the time being, however, household arrangements are context-specific. Households adjust their arrangements to the possibilities and constraints of a given context. Within the possibilities and the limits of the housing market, households shift over the life-course between urban and suburban contexts to secure a supportive environment.

REFERENCES

Clark, S. D. (1966) *The Suburban Society*, Toronto: University of Toronto Press.
Cook, C. (1988) 'Components of neighbourhood satisfaction: responses from

urban and suburban single-parent women', *Environment and Behavior* 20: 115–49.

Fagnani, J. (1993) 'Life course and space: dual careers and residential mobility among upper-middle-class families in the Ile-de-France region', in C. Katz and J. Monk (eds) *Full Circles: Geographies of Women over the Life Course*, London: Routledge.

Fava, S. F. (1988) 'Residential preferences in the suburban area: a new look?', in W. van Vliet (ed.) *Women, Housing and Community*, Aldershot: Avebury.

McDowell, L. (1983) 'Towards an understanding of the gender division of urban space', *Environment and Planning D: Society and Space* 1: 59–72.

MacKenzie, S. (1989) 'Restructuring the relations of work and life: women as environmental actors, feminism as geographical analysis', in A. Kobayashi and S. MacKenzie (eds) *Remaking Human Geography*, Boston/London: Unwin Hyman.

Michelson, W. (1977) *Environmental Choice, Human Behavior and Residential Satisfaction*, New York: Oxford University Press.

Organization for Economic Co-operation and Development (1993) *Employment Outlook*, July, Paris: OECD.

Popenoe, D. (1977) *The Suburban Environment: Sweden and the United States*, Chicago: University of Chicago Press.

Pott-Buter, H. (1993) *Facts and Fairy Tales about Female Labour, Family and Fertility: A Seven-Country Comparison, 1850–1890*, Amsterdam: Amsterdam University Press.

Pratt, G. and Hanson, S. (1991) 'On the links between home and work: family-household strategies in a buoyant labour market', *International Journal of Urban and Regional Research* 15: 55–74.

Rapoport, R. and Rapoport, R. N. (1969) 'The dual-career family', *Human Relations* 22: 3–30.

Riesman, D. (1958) 'The suburban sadness', in W. Dobriner (ed.) *The Suburban Community*, New York: Putnam.

Rose, D. and Chicoine, N. (1991) 'Access to school day care services: class, family, ethnicity and space in Montreal's old and new inner city', *Geoforum* 22: 185–201.

Rossi, P. H. (1955) *Why Families Move: A Study in the Social Psychology of Urban Residential Mobility*, Glencoe, IL: Free Press.

Rothblatt, D. N., Garr, D. J. and Sprague, J. (1979) *The Suburban Environment and Women*, New York: Praeger.

Saegert, S. (1980) 'Masculine cities and feminine suburbs: polarized ideas, contradictory realities', in C. R. Stimpson *et al.* (eds) *Women and the American City*, Chicago: University of Chicago Press.

—— (1988) 'The androgynous city: from critique to practice', in W. van Vliet (ed.) *Women, Housing and Community*, Aldershot: Avebury.

Wekerle, G. (1984) 'A woman's place is in the city', *Antipode* 16: 11–19.

Whyte, W. H. (1956) *The Organization Man*, Garden City, NY: Doubleday Anchor Books.

12

FAMILY, GENDER AND URBAN LIFE: STABILITY AND CHANGE IN A COPENHAGEN NEIGHBOURHOOD

Kirsten Simonsen

INTRODUCTION

This chapter addresses the development of the mode of life of Danish women and their families based on a case study of a working-class neighbourhood in Copenhagen. The period under examination covers rather more than fifty years, from the building of the neighbourhood in the 1920s until 1990 when the study was conducted. Its goals are to explore the areas of continuity and change in the family and the neighbourhood – in social interaction and in attitudes towards work, seeing these through the eyes of the residents but in such a way as to understand the interplay between individual biographies and larger social structures.

In Denmark as a whole, this period has been one of tremendous changes in the conditions of women's lives in general and in women's work in particular. (For a closer analysis of these developments see Simonsen 1990.) During the 1920s and 1930s, women increasingly left occupations in domestic service and industrial home-work in favour of workplaces spatially separated from the home – in manufacturing industries, service industries and social services. After the Second World War, however, the ideal of the housewife at home gradually became a reality for the working class, not only the middle class, as the general improvement in welfare made it economically possible for an increasing number of married women to remain at home full-time. Their numbers grew in the 1950s and peaked in 1960 when 48 per cent of all women between 14 and 69 years of age were housewives. Accompanying this development was a 'home economics' movement, which elevated the social status of women's domestic labour. The dominant discourse was a 'separate but equal' one which reinforced the home and the local community as women's spaces.

The postwar years saw a related decrease in participation in the labour force by both married and single women but the picture shifted in the 1960s and the rate of female participation has been growing ever since (see Figure 12.1). The major single employer responsible for this growth has been the expanding welfare state and especially the services that support social reproduction. Since the 1970s, about half of the female labour force has been employed within the public sector.

Connected to this development is a change in the provision of child-care facilities that is both quantitative and qualitative. Before the 1960s, kindergartens and other children's institutions were primarily seen either as social assistance arrangements aimed at taking care of children of working-class women forced to engage in wage-labour or as a part-time supplement to child-minding at home. In the 1960s, a higher priority was given to the educational purposes of childcare institutions and they came to address themselves to children from all classes. This change led to an increase in the coverage by kindergartens (age 3–6 years) from 9.4 places per 100 children in 1961 to 67 per 100 children in 1991. For day nurseries (0–2 years), the percentages are somewhat lower, reaching 48 per cent in 1991. The staffing of these centres provided new employment

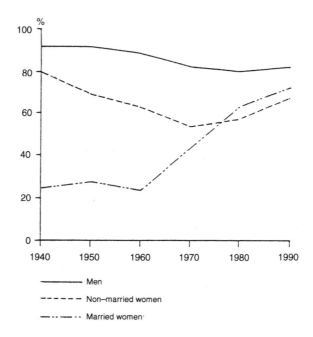

Figure 12.1 Activity rates for men, non-married women and married women, 1940–90

Sources 'Living Conditions in Denmark'. Compendium of Statistics. Statistics Denmark and Institute for Social Research, 1992

opportunities for women while the provision of childcare made it possible for high proportions of women to combine paid work more readily with their domestic responsibilities.

These are some of the social changes that set the context for the development of women's modes of life in a neighbourhood of apartment blocks in the central part of Copenhagen, a choice implying what I, in a broad sense, consider a working-class population. The investigation employs in-depth qualitative interviews with the women concerning their life-histories and those of their families.

THE LIFE-HISTORY BETWEEN STRUCTURE AND PRACTICE

The analytical method chosen for this research is strongly related to the theoretical understanding involved (see Simonsen 1991). On the level of social ontology it means that the method has to support the maintenance of what Giddens (1979, 1984) has termed the duality of structure – that is, it has to match a theoretical approach claiming that neither a structural-determinist nor a voluntarist approach is satisfactory. According to the notion of duality of structure, structures are constituted by human practices, but at the same time they are the very medium of this constitution. They are not only sources of constraints, but also the enabling media of action. This allows, in some sense, a view of human beings as capable and knowing agents who know what they do when they do it.

On the level of epistemology, the method should relate itself to what Giddens calls the *double hermeneutic* of social science – that is, the intersection of two frames of meaning which are a logically necessary part of social science: the meaningful social world as constituted by lay actors and the metalanguages invented by social scientists. Social analysis is constantly involved in a process of interpretation or translation between the two.

The life-history interview can be characterized as a variant of the in-depth, qualitative interview having the life-courses – or particular aspects of them – of individuals or groups as their object. Theoretically, it is connected with the relationship between individual practices and sociohistorical change. In this relation, the human being cannot be seen as a particular individual exclusively, but rather as a Sartrean 'universal singular'. According to this notion, ordinary people universalize, through their lives and their actions, the historical epoch in which they live. They are singular instances of the 'universality of human history' (Sartre 1985). That means that their norms and their practice are currently produced and marked by personal as well as social history.

In the interviews, *social practices* and the meanings given to them by individuals themselves are the object of concern rather than attitudes and psychological individuality. Likewise, emphasis is placed on turning-points and pronounced events in the life-courses of the individuals, following the point of view that such incidents provide the best openings towards their norms and life-strategies. This practice also reveals the interrelationship between the various practices and life-spheres of the individuals and their families. What is needed in this connection is a multidimensional concept of the life-course (work, consumption, parenthood, marriage, etc.) that represents the interdependent careers of the different members of the families. Godard (1981) offers such a concept in talking about 'familial conjunctures of practices', the conjunctures referring to the particular situation into which the practices are inscribed both in relation to the life-course of the families themselves and to social history at different levels; the familial to the entirety of practices – often arranged into sequences of actions – which constitute the situation of the family. In this way, the life-course widens from being an individual matter to being a familial one as well.

The temporal aspect is an essential part of the life-history interview. Through a repeated movement 'back-and-forth' between the biography and the epoch, it renders possible at the same time understanding of the specific individual and the social setting. The biography exposes the specific socialization of the individual during the life-course, particularly during childhood and the period of establishment. In this regard, the epoch determines the field of possibilities, the field of instruments and so on. Furthermore, the socialization and the life-course of the individual subject grow out of not only a temporal but also a spatial setting. The locality of birth and adolescence, the trajectory of residence, and the 'locales' of day-to-day interaction are integral parts of the biography. All things considered then, the situated life-history provides a useful methodological entrée to a contextual understanding of mode of life and to a search for mediation between structure and human practice.

In the concrete analysis, these considerations, together with the initial purposes, result in a research design that involves four basic analytical perspectives.

1 *The perspective of class* represents an intention of exposing a broadly defined contemporary urban worker's life-world. In the segregated city, the choice of neighbourhood for analysis determines its class dimension; in this case the neighbourhood chosen consists of apartment blocks from the middle of the 1930s, physically in good condition but still available at a moderate rent.

2 *The perspective of family and gender* represents the fact that the bearers of a mode of life will be the primary group as much as the singular individual. In this view, relations between the members of the families are an integral part of the mode of life and the starting-point in the life-histories of the women is intended to bring these relations into focus.

3 *The perspective of locality* represents the concept of spatiality in the

mediation between structure and practice (Simonsen 1991). The analysis explores the importance of residential location as well as the meaning of the neighbourhood and/or the city in the mode of life, together giving the study a dimension of place.

4 *The perspective of generation* represents the concept of temporality in the same role. In the choice of interviewees, four generations of women are represented so that change and stability in the mode of life during the fifty years of existence of the neighbourhood are exposed through intra- and inter-generational variations.

HOW THE LIFE IS FORMED

The life-histories of the responding women and their families are formed in a field of interaction between the unique and the universal, between the specific human relations in which they have been involved during their life-course and the social conditions dominating their generation and their locality. At the same time, however, the life-histories reveal a number of common features cutting across the different individuals, families and generations. Together these features, interconnected as they are, produce a general image of the mode of life of the families and, in some sense, of the neighbourhood as well.

The family

The Marxists of today only worry about the grown-ups: reading them, you think that we were born at the age where we gain our first salary; they have forgotten their old childhood and everything happens, reading them, as if the human beings feel their alienation and their reification *first in their own work* while everyone actually experiences it first as a child, *in the work of their parents.* On the contrary, Existentialism thinks it can integrate this method, because it finds the insertion point of individuals into their class, that is, the single family as mediation between the universal class and the individual. (Sartre 1985: 57; my translation)

With these comments, Sartre identifies the family as a *mediation category* between structure and agency, and childhood and adolescence as the periods in which the basis of the life and the personality is formed. Adolescence is the primary period of insertion into social relations (class relations, gender relations, place-based relations, etc.), so that the temporal and spatial contexts in which this period is experienced are of great importance to the life-course as a whole. This importance, in a very concrete manner, can be illustrated through the Copenhagen life-histories. For instance, a common feature cutting across the stories of the individual families is that at least one adult member of the family has grown up in the neighbourhood, or in close proximity to it, and this, obviously, has been a basis for a sense of belonging and mutual connection between kin and neighbours. Another example is a tendency in many families to reproduce occupational status from one generation to another, mostly in the form of the man being a skilled worker and the woman a clerk in the public or private sector.

The mediating role of the family thus functions as a source of reproduction or stabilization of the mode of life of the neighbourhood. Additionally, the role of the family is sufficiently important to justify the designation of a *family-orientated mode of life*. Family life, and particularly raising children, is given high priority in everyday life as well as through the life-course. 'Marry and have a family' exists as a 'quasi-natural' value in the mode of life, even if formal marriage itself is not of vital importance any more, and a good many family resources are invested in the raising of children. Furthermore, some normative impulse exists to establish your family at a relatively early age in order to 'be young with your children'. Listen to the priorities as they are expressed by some of the women:

> So we have been incredibly sensible, you can say. I was working part-time and Ib was unemployed, and still we didn't miss anything. Because we were careful about how to use the money. But what we constantly have in view is that the kids – as far as we can manage – have a safe childhood. (Vibeke, 40 years old; my translation)

> We take part a lot in the parents' activities at the school, and when, earlier, Soren attended day nursery and kindergarten, we were also strongly involved in that. I mean, it is very important that we concentrate on the welfare of the children – that you can take part in everything. (Inge, 40 years old; my translation)

In some sense, this family-orientated mode of life is counteracted by the process of individualization in modern life which can be described as an increasing tendency for an individual to be integrated into non-familial activities, organizations and institutions, only representing him- or herself. Examples are work, union activities and leisure activities such as education and sports. In the mode of life, this contradiction comes through as a *normative duality*. On the one hand, family life and family activities are the centre of life, on the other, the women emphasize having their own activities and their own friends and relationships outside the family circle.

> In the years when I was a housewife, I took classes in typing and joined some evening courses – maths and Danish and gymnastics . . . I was not just at home all the time only seeing my husband and children and the closest family. I kept contact with people with whom I had joined courses and I had my own social activities. Actually, I have always been doing that. Even if I have been a housewife many

years, I have always been involved with people independent of my own little nuclear family. (Jytte, 50 years old; my translation)

A major thesis in all modern family theory is that the extended family of the past, consisting not only of parents and their children but also of uncles and aunts, nephews and nieces, cousins and grandparents, has shrunk into a smaller body, and that the nuclear family has very loose contacts with other kin who are spatially dispersed all over the country. The descriptions of the lives of the Copenhagen families given thus far could support this thesis. However, even if the focus on the nuclear family is strong, most of the women, in all the generations represented, also refer to a *close kinship network*. Many family members have kin in or in close proximity to the neighbourhood and they see them or talk to them quite frequently.

I wouldn't think about moving. It is because I am very close to my family and many of them live quite close to us. Our kids, of course, but also some of my sisters and brothers . . . Just the thought that I can pop up to them, maybe only for an hour or so, means a lot to me. (Jytte, 50 years old; my translation)

Yes, very close. Many don't understand it, but we are incredibly close. I do think it has something to do with living so close to each other. The whole family together, we are just enjoying ourselves. (Lisbeth, 25 years old; my translation)

The kinship network not only expresses itself in frequent contacts, it is also the basis of more binding modes of practice such as mutual care and assistance among generations. In this connection, the most common type of service is child care, which has been of particular importance for those who are single parents.

I have been lucky that my parents live at the other side of the street, they have been very helpful. My daughters never went to day nursery, none of them did. They attended kindergarten when they became 3 years old. But before that my mother wanted to look after them, and she really did it well I can assure you. (Mette, 55 years old; my translation)

But other activities, such as helping elderly parents or giving room to young couples without a residence, have also been mentioned:

For seven years, my mother couldn't leave her home and needed help. She didn't want public help – because she had me! So every day after work I went up there to see what she needed from town and things like that. She lived just around the corner. But Mickey, our son, he started shopping for her when he was 14 years old. He did that until he was apprenticed at the age of 18. Sometimes I thought 'It is incredible that it works.' He tells his friends: 'I can't come now, I have to do some shopping for my grandmother.' (Laila, 45 years old; my translation)

Given such expressions, it is reasonable to talk about an element of the 'extended family' in the mode of life of the neighbourhood.

Neighbourhood attachment

As already indirectly illustrated, a close connection exists between kinship networks and *neighbourhood attachment*. A good many family members have their parents or their children in the neighbourhood and even have brothers and sisters, aunts and uncles and cousins as well; these connections naturally provide a basis for neighbourhood attachment. However, other kinds of relationships can also form the basis of attachment to people and place. John Eyles (1985) in his book *Senses of Place* identifies three interrelated and irreducible elements of 'community' or 'place' which can be recognized in the present case as well. Eyles relates the three dimensions to a discussion of the concept of community, but I will not take up that aspect here, both because of the value-claims involved in the concept and because it implicitly presents the idea of neighbourhoods as closed social systems, an irrelevant perspective for interpreting modern urban life. The three dimensions in this case, then, figure exclusively as possible elements of an attachment to place.

First, the ecological dimension is based on *geographical area*. It sees the locality – the arena of activities, institutions and practices – as important in its own right. In adopting this view, Eyles relates to a discussion of the ecological determinism of the Chicago School, but it can be seen as well simply as the attachment of the individual to her material and functional *environment*. Concretely, the environmental factors accentuated by the inhabitants of the Copenhagen neighbourhood express a double interest of family life in an urban context. The women speak of having a nice and quiet neighbourhood which is 'good for the children' and which at the same time offers good accessibility to all kinds of urban facilities:

> I can't even imagine us moving from this place. Because everything is very easy. It is nice and quiet for the children, that is, a quiet neighbourhood. Here are nice green areas, and still you are in the middle of the city. This is nice, I think, and easy access to everything – bus and train and things like that. (Bente, 30 years old; my translation)

The second dimension of place, social structure or *social networks*, is linked to interactions, relationships and institutional practices referable to the locality. The families interviewed express general agreement on this dimension. Everyone knows everyone else, maybe not by name, but at least by face. They know each other from the streets and lawns between the blocks, from the local shops and the laundry and from the co-operative housing associations. Or they know each other from school days, reflecting the fact that everybody who has grown up in the neighbourhood has attended the same school, or from the institutions attended by their children where parent associations form a new area of contact. These networks also form a basis of mutual aid and they are very important to the inhabitants:

> Well, if the situation was that I were forced to move, I could put up with that. Or rather, you would have to do it, wouldn't you? But I think I would miss this place, because just to walk on the street and meet somebody and 'Hey, how are you? How is everything?' and so on, that is very nice. It is like living in a small town. (Connie, 60 years old; my translation)

> So you know them. You know who they are and approximately where they live. And because I have lived on Glaswej as well, I also know a lot of people from there, and from the children's institutions. Before school they have been in both day nursery and kindergarten and you get to know a lot of people through those institutions. You can take a walk without meeting somebody you know, but it rarely happens. There is always somebody you know and to whom you just say 'hello' – and I like that. (Vibeke, 40 years old; my translation)

Finally, the third dimension refers to place as an ideological structure, in this case defined as *sense of belonging*. This dimension, which is about sentiment and commitment, may exist beyond the level of consciousness, implicitly embedded in social practice. However, it comes to the surface, for instance, when a woman says: 'I couldn't even think about moving from this place. You see, this neighbourhood, this is my home' (Laila, 45 years old); or another: 'I don't dare move from here, because I will just move back again! You are always seeking to come back again – really I don't know why!' (Karin, 35 years old). These people cannot explain their attachment to the neighbourhood. They simply belong. Place is an integral part of their mode of life.

While all three dimensions of place are involved in neighbourhood attachment, the social dimension evidently is the dominant one. Most of the respondents express attachment to the neighbourhood on the basis of social ties to kin, friends, or neighbours. That does not mean, however, as is implicitly assumed in the community debate, that the neighbourhood is a closed social system where the families have the major part of their social involvement. The social networks of the households are much more spatially differentiated and they also link to employment and leisure activities all over the city. But they nevertheless express a neighbourhood identity based on social ties between the inhabitants. The interesting question, then, is about the character of these social ties. Let us take an example:

> Everybody living on this staircase, we all talk together, but we don't camp on each other's doorstep. We drop in occasionally, but it is not

very usual. We know everybody here and talk a lot to them, but we don't associate privately. (Laila, 45 years old; my translation)

Besides the kinship networks and the few close friends in the neighbourhood which some of the people interviewed have acquired, most of the social interaction between the residents in the neighbourhood consists of *relatively* weak ties. You say 'hello' in the street, you talk together, you meet on the lawns and you have a summer party together, but you do not have relations of a primary type with your neighbours. The importance of such weak ties should not be underestimated, however. As recent research on networks has shown, they serve important needs (see, for example, Schiefloe 1990). They allow the individual to engage in meaningful interactions with a wider community than her immediate circle and they open possibilities for useful exchanges of goods and services. At the same time, weak ties can be kept under control with regard to obligations and intrusions of privacy. As Fischer (1981) has argued, widespread networks of weak ties in a neighbourhood make it safe and friendly, turning it into an identifiable 'private world'.

Work and interaction

Together with the family (both the nuclear family and what can be termed the extended family) and the neighbourhood, work emerges as an important producer of identity in the life-histories of most of the women interviewed. Some of them were full-time housewives for a shorter or longer period of their lives, but most of them have been in wage labour the major part of their adult existence. Today, women's education and participation in the labour force are integral parts of the mode of life and these activities are considered in the best interest of families, since both men and women need the stimulus from the outside:

I didn't want to be at home – not at all! Actually, I think it is only a few who want to stay at home. Seeing the man coming home and a lot of things have happened to him and you just have been between nappies at home – nothing has happened at all. You don't have anything to talk about – it is depressing. I think it is nice that you each have some experiences which you can talk about and agree or disagree about. (Laila, 45 years old; my translation)

As a time-space structured activity, this participation, obviously, has decisive consequences for the *routinization* of everyday life of the families. Let us pick out an example:

Well, Allan, he gets up in the middle of the night – a little past 6.00 or something like that. He gets up quite early, I mean, and goes to the baker's, and makes coffee, and reads the paper. Then I get up about 7.00, and the coffee and the roll are ready, and then I am

reading the paper until a little before 8.00. Allan leaves for work at
7.45 and I leave a little past – I start work at 8.30. Then each of us
works and Allan normally is home between 5.00 and half past, after
which he does the dishes and tidies up a bit. Then I am back
between 5.00 and 6.00 and I usually cook, and then we eat. And
then I sometimes leave for some kind of meeting. Or, if we stay at
home, one of our kids may drop in during the evening. (Jytte, 50
years old; my translation)

The twofold connection to the labour market makes workplaces the
social institutions most decisive to the structuration of everyday life,
and the other routines of everyday life, such as the division of labour
in the home, are adapted to this predetermined activity.

Employment, then, is a major element of structuration in everyday
life, in the life-course and in the personal identity of both male and
female family members, not only as a means of obtaining a salary, but as
an activity playing a part in the individual's system of meaning. But
which are the qualities of employment that make it an important
dimension of life and identity? In discussing that question, the women,
whatever jobs they held – in manufacturing industries, in social services,
or, most commonly, in some kind of office work – concurrently emphasized interaction or personal contact with other people as the most
important dimension of work, whether with customers and clients or,
particularly, their colleagues. Around the latter group a sense of collec-
tivity, or more precisely a sense of being part of a 'we', is developed as a
work-based contribution to identity formation.

I don't even look first at the salary. I look at the colleagues, if you
have a nice atmosphere and can talk to each other. That is of
primary importance to me – and always has been so . . . So we
are sitting in a room. We are five women and one who sets the tools
for us. I am 45 years old, and there is one of 47, one of 49 and one of
55, and then there is one of 36. She is quite a lady, she keeps things
going. So we always have a lot of fun – not a day ends in which we
haven't had a good laugh. (Laila, 45 years old; my translation)

This finding, together with the previous ones, stresses the importance
of work at the same time that it questions the Marxist tendency to regard
work as the only source of identity formation. A wider approach sees
membership in various *groups* as the constitutive element of individual life
and as the mediator between structure and practice. In this way, the
group becomes the privileged domain of that singular universal which
methodologically is stressed as decisive to the handling of the problem of
mediation (see also Ferrarotti 1981; Sartre 1985). An argument for a
concept of practice related to this point of view was given years ago by
Habermas in a critique of Marx for his reduction of the duality of social
practice. In Habermas's view, a concept of social practice must include

both *work and interaction*: on the one hand the instrumental action regulating the material exchange of the human being with nature, and on the other the communicative action which is about the relationship of intersubjectivity between social individuals (Habermas 1968a; Habermas 1968b).

Life-strategies

One theoretical starting-point of this analysis was the recognition of human beings as knowledgeable agents who know what they do while they do it, a recognition that implies a concern with human intentionality and motivation to the extent to which these turn up in the life-histories. One way of analysing intentionality and motivation, using a notion introduced by Cuturello and Godard (1982), is to examine the ways in which individuals and families *mobilize* their resources during their life-course in order to attain particular goals. An analysis of the life-histories from these perspectives, however, yielded only a few instances of conscious mobilization for the achievement of particular goals. One good example is a family in which husband and wife have mobilized their resources – materially, financially and morally – around his achievement of a new kind of job through higher education, and in which they now discuss the possibility of improving her qualifications as well.

Even if most of the families do not perform such conscious mobilizations, it is not that they have no priorities or strategies for their lives. As the discussion of the family-orientated mode of life revealed, children are put high on the list of priorities, not so much with respect to goals concerning their education or their future, but rather with the objective of giving them a childhood which is nice and safe.

> Still, we did get some material things. But our priority is as far as possible to give the children a good childhood . . . And then we agreed that our second child, she became our investment. At the time when people bought cars and other things, we decided to have another child – for she uses the money we could have used to get other things. (Vibeke, 40 years old; my translation)

More generally, the life-strategies of the families can be characterized as focusing on *human relations* – first and foremost, relations to the closest family, but also to the extended family, the neighbourhood group, and the group of co-workers, and as emphasizing safety and security. Families aim to produce a financial safety-net which can stand up to unforeseen events such as unemployment and illness. Part of the strategy, also, is a desire to create a freedom and possibility of choice for themselves in relation to work, marriage and so on, often resulting from a reflexive evaluation of the possibilities of choice of the former generation as being very limited.

In a way you can say that we have acted purposefully – not for something material, but simply for the freedom to stop if anything unexpected should happen. The only thing we actually have kept in view all the time is securing the possibility that one of us could stop working if the work got too rough or if something happened to one of our kids or other members of the family. (Jytte, 50 years old; my translation)

At the same time, however, the content of the strategies is marked by the class background and the life-situation of the families and is adapted to the historical conditions and horizons of possibility. This finding points to the importance of discussing the concept of *strategy* itself. In this connection, I consider the concept as it is advanced by Bourdieu (1977, 1990) a useful approximation. Bourdieu starts his discussion of strategies by focusing on the concept of future. He rejects a Hegelian 'absolute possibility' (*absolut Möglichkeit*) in favour of a probable 'upcoming future' (*un à venir*). That is, the individuals do not act in relation to any future that may occur, but rather in relation to a future which is connected with an estimation of the available possibilities, a system of objective potentialities immediately inscribed into the present through the schemes of perception of the individuals themselves.

In this sense, the concept of strategy is closely connected to one of Bourdieu's most central concepts: the concept of *habitus*, which is his answer to the problem of mediation between structure and practice. Habitus is a system of dispositions consisting of cognitive and motivating structures predisposed to function as principles which generate and organize practices. Habitus is the product of history as well as biography, an internalization of externalities, not in a mechanical way but rather as a limitation on the thinkable. The non-possible is excluded from the consciousness as non-thinkable. Habitus is a quality of individuals, based on the singularity of their social trajectories, but the habitus of members of the same group can be united in a relationship of homology, that is, of diversity within homogeneity, reflecting homogeneous living conditions that are creating common schemes of perception, conception and action.

An application of this approach to Copenhagen life-histories at first underlines that the individuals and the families do have schemes of actions and priorities, for example, concerning the extended family, attachment to place and patterns of consumption which cannot be derived in a mechanical way from modern, capitalist and urban development. At the same time, however, we observe a tendency for families' schemes of action and priorities to reproduce the life-situation of which they are formed. For instance, when several women claim that the kind of job in which they landed was accidental, the actual choices show the influences of their own dispositions or habitus and of the social networks to which they have access. Similarly, the families' life-strategies actually aim to sustain the social relationships of which the families form a part and protect them against outside disturbances. Bourdieu puts this process as follows:

> Through the systematic 'choices' it makes among the places, events and people that might be frequented, the habitus tends to protect itself from crisis and critical challenges by providing itself with a milieu to which it is as pre-adapted as possible, that is, a relatively constant universe of situations tending to reinforce its dispositions by offering the market most favourable to its products. (Bourdieu 1990: 61)

STABILITY AND CHANGE IN THE MODE OF LIFE

Cultural writings have characterized life in a modern urban context as rootless and flexible; thus, for instance, Berman (1982) talks about how 'all that is solid melts into air'. Yet a striking characteristic of the mode of life is its relatively large element of *stability* in social life and cultural meaning. This stability stands out in the residential trajectories of individuals and families whose members have stayed in the neighbourhood a great part of their life-course, and who have parents, children, or other kin nearby as well. In this context, the time of residence and the kinship networks are major contributors to the stability of the mode of life. But at the level of meaning as well, stability exists, since values and priorities such as family orientation, neighbourhood attachment and a certain rejection of the dominant consumer culture appear as common ballast across the generations. Except with respect to material wealth, the respondents did not experience great difference between their own lives and those of their parents or children. It can thus be argued that an extensive cultural transmission is functioning between the various generations in the neighbourhood. This does not mean, however, that diverse modes of life or less stable residents do not exist in the neighbourhood, but only that this family-orientated and place-attached mode of life is sufficiently important to build a social milieu there.

At the site of employment, by contrast, the tendency is for development from a very stable situation in which the residents have a lifelong connection with a single place of work towards a situation marked by frequent changes and flexibility, in all probability as a consequence of the social changes that are accompanying the economic restructuring that is occurring in industrial societies. Thus we see the contrasting expressions of two women:

> I think, I have taken root in this work. I don't think I would fancy anything else. I don't know, you probably can't tell when you have been in the same firm for nearly forty years! You must like it, or you wouldn't stick around that long. And it has been fun to participate in

the development – I really do think so. (Mette, 55 years old; my translation)

Then, it is about trying something else. I have realized that I actually like that. I have never been in one place of work for many years. It is my subconsciousness that works and says: 'Now you have had it, you know the walls too well, you don't feel like it any more. Now something has to happen.' (Vibeke, 40 years old; my translation)

This development in workplace culture, however, seems partly to be compensated for by the steady emphasis on family and neighbourhood which makes these spheres the locus for the creation of the safety and the security needed in social life.

While the pronounced stability in residence and life-strategy across the generations can be interpreted in accordance with the tendency of the habitus to reproduce the life-situation by which it has been formed, one sphere of life has experienced important changes in normativity and practice closely connected with an overall change in social ideology in Danish society – the sphere of *gender relations*. First and foremost, these changes manifest themselves in relation to the question of women's participation in the labour force. In the oldest generations, the dominant value is that the most favourable situation for the family and the children is for the woman to be a housewife, not a worker away from home. This view is upheld no matter whether the norm is in accordance with the actual practice of the women. A good example is that of an old woman who has been running a small shop in the neighbourhood during most of her marriage and who is very hostile towards public day-care. In her self-image she manages to overcome ideological contradictions by interpreting the shopkeeper and the mother role as a whole. She never really regarded herself as being away from home. In the younger generations, on the contrary, women's having employment is ideologically taken for granted, any deviation from that situation calling for justification.

The other changes in gender relations are connected with this primary one. One involves the attitude towards the gender division of labour in the home. Amongst the oldest women you can find a full acceptance of the man's not doing anything at all in the home, notwithstanding both man and woman having full-time employment. In the other generations the value-claim is that 'you share the duties'. This does not indicate, by any means, that the actual gender division of labour has been equalized, but it shows that it has become an area of negotiation which most of the women described as low-conflict. The exception is a Danish-Irish couple in which the difference in cultural traditions in relation to gender roles gives occasion for conflict.

Another sphere in which the influence of changes in women's participation in the labour market has been felt is in the system of social networks in the neighbourhood. Up to about 1960 the local social

networks were based to a great extent on the children and the contacts
between mothers working at home; they were established in connection
with childcare, shopping, laundry and so on. This situation was supported by the existence of a range of small shops in the local neighbourhood. With the death of small shops, the disappearance of housewives,
and the departure of children to their institutions, these kinds of contacts
diminish, and so does the daytime liveliness of the neighbourhood.
Other kinds of personal contacts appear as a new basis for networks,
however, and they involve both men and women to a much greater
extent. Again, the children are the starting-point; parents' organizations
around day-nurseries, kindergartens, schools and children's leisure activities are the places in which you meet other parents today. Additionally,
contemporary social networks are based on more-or-less organized
common adult activities such as participation in housing associations
and common use of the semi-public spaces of the neighbourhood for
daily recreation, summer parties and so on.

Probably this development of the gender relations is the major factor
distinguishing life in the neighbourhood today from that described as the
traditional, urban, working-class mode of life. The classic analysis by
Young and Willmott in Bethnal Green in East London (1957) describes a
mode of life involving a strong neighbourhood attachment based on
local communities of kin and neighbours. The dominant figure of these
communities was the 'Mum', and the woman's relationship was most of
all with her mother, but also with her other female kin – these relationships formed the basis of local social networks. In adolescence, Young
and Willmott write, the daughter seeks the same freedom from parental
influence as her brother, and as long as she is working she behaves like a
man.

> But when she marries, and even more when she leaves work to have
> children, she returns to the woman's world, and to her mother.
> Marriage divides the sexes into their distinctive roles, and so
> strengthens the relationship between the daughter and the mother
> who has been through it all before. (Young and Willmott 1957: 61)

Neighbourhood life, then, was based on strongly gender-segregated
networks. The mother-centred women's network provided working-class
women with some security in life in case of desertion by their husbands
but also broke the sense of isolation associated with childcare and other
reproductive activities. In the case of the men, communities were sought
outside the family in relation to the place of work or in the public life of
the pubs.

Similar features have been recognized in more recent investigations of
the mode of life as well. Bleitrach and Chenu (1979), for instance, in
their analytical delineation of an urban, skilled workers' mode of life in
the south of France, emphasize the interrelated familial and local
organization as well. The families they analysed are organized with a

strong, traditional gender division of labour and they are part of a local social life which is organized in sharply gender-segregated social networks. Everyday male communities cut across the spheres of the family, work, leisure time and union activities; by contrast, as above, female networks concentrated on housekeeping and reproduction activities. Similarly, Andersen *et al.* (1985), studying a late nineteenth-century quarter in central Copenhagen, still recognize elements of local attachment based on 'matrilocalization' and women's networks organized around the mother-daughter relationship.

While the present analysis clearly recognizes the neighbourhood attachment and family networks, elements of a gender-segregated social life have been traceable only in the life-histories of the very oldest inhabitants and in the childhood descriptions given by a few of the women. In some sense, then, the mode of life I have described can be interpreted as a *neoculturation* of the traditional, urban working-class mode of life encouraged by the social development of gender relations.

CONCLUDING COMMENTS

Obviously, the description and the interpretation given in this chapter of the mode of life of women and their families in a specific neighbourhood in Copenhagen should not be seen as a depiction of an understanding of the urban way of life in the contemporary European city. It is a case study analysing the mode of life that exists in a specific urban context at a specific period of time. It should be seen as an example illustrating the great diversity of modes of life coexisting and confronting each other in the late modern city. Still, a few general conclusions can be drawn from the case.

First, the analysis supports the warning against overall generalizations and presuppositions as to the lives and social practices of people across time and space. Two widespread generalizations are challenged by the findings of this case. The one is the general understanding of modern urban life as totally disembedded with respect to place attachment, family and kinship networks and primary social relations. The elements of neighbourhood attachment, social networks and cultural transmission between generations found in the analysis underline the importance of recognizing continuity as well as breaks in the understanding of contemporary urban social life. The other presupposition challenged is about the differences between northern and southern parts of Europe. One theme which is often presented as a major difference concerns the existence of close family and kinship networks in the south as opposed to the north. In this regard, the analysis shows that the family and kinship networks can be found even in the cities of a society as penetrated by welfare institutions as the Danish one, even if their practical importance in the management of day-to-day life is probably more limited.

Second, the results of the analysis underline the importance of the development of gender relations for the understanding of urban social life. The thesis of women as a secondary labour force – dependent on men and having a possible retreat to more 'traditional' activities – on which much urban and social theory is built, is now totally untenable in most European contexts. It is based on a lack of understanding of the substance of women's participation in the labour market. Especially in countries such as Denmark, where female participation in paid employment has become the norm, wage labour has become an important element in contemporary female identity with implications for other spheres of life as well.

REFERENCES

Andersen, L., Ravn, P. and Thomsen, M. (1985) 'Et kvarter i familiens sk d – ommdreogd tre på broerne', unpublished MA thesis, Roskilde University.

Berman, M. (1982) *All That is Solid Melts into Air: The Experience of Modernity*, New York: Simon & Schuster.

Bleitrach, D. and Chenu, A. (1979) *L'Usine et la vie. Luttes régionales: Marseille et Fos*, Paris: François Maspéro.

Bourdieu, P. (1977) *Outline of a Theory of Practice*, Cambridge: Cambridge University Press.

—— (1990) *The Logic of Practice*, Cambridge: Polity Press.

Cuturello, P. and Godard, F. (1982) *Familles mobilisées*, collection, texte intégral, Paris: Plan Construction.

Eyles, J. (1985) *Senses of Place*, Warrington: Silverbrook Press.

Ferrarotti, F. (1981) 'On the autonomy of the biographical method', in D. Bertaux (ed.) *Biography and Society*, Beverly Hills: Sage Publications.

Fischer, C. S. (1981) 'The public and private world of city life', *American Sociological Review* 46: 396–416.

Giddens, A. (1979) *Central Problems in Social Theory*, London: Macmillan.

—— (1984) *The Constitution of Society*, Cambridge: Polity Press.

Godard, F. (1981) *A propos de pratiques familiales. Contribution à l'approche anthroponomique*, Nice: Faculté de lettres et sciences humaines, Université de Nice.

Habermas, J. (1968a) *Erkenntnis und Interesse*, Frankfurt am Main: Suhrkamp.

—— (1968b) 'Arbeit und Interaktion. Bemerkungen zur Hegels Jenenser "Philosophie des Geistes"', in *Technik und Wissenschaft als 'Ideologie'*, Frankfurt am Main: Suhrkamp.

Hojrup, T. (1983) 'The concept of life-mode. A form-specifying mode of analysis applied to contemporary western Europe', *Ethnologica Scandinavica*: 15–50.

Sartre, J.-P. (1985) *Critique de la Raison dialectique – précédé de Question de méthode*, Paris: Edition Gallimard.

Schiefloe, P. M. (1990) 'Networks in urban neighborhoods: lost, saved or liberated communities?', *Scandinavian Housing and Planning Research* 7: 93–103.

Simonsen, K. (1990) 'Urban division of space: gender category!', *Scandinavian Housing and Planning Research* 7: 143–53.

—— (1991) 'Towards an understanding of the contextuality of mode of life', *Environment and Planning D, Society and Space* 9: 417–31.

Young, M. and Willmott, P. (1957) *Family and Kinship in East London*, Harmondsworth: Penguin Books.

13

REGIONAL WELFARE POLICIES AND WOMEN'S AGRICULTURAL LABOUR IN SOUTHERN SPAIN

Maria Dolors García-Ramon and Josefina Cruz

This chapter focuses on two themes: first, the daily life and work of women agricultural day-labourers, a conspicuous group in southern European countries, and second, the ways in which welfare policies affect the everyday lives and work of these women. Despite the emergence of feminist research on the role of women in the agricultural sector in developed countries over the last decade (Barthez 1982; Canoves *et al.* 1989; García-Ramon 1990; Haugen 1990; Sachs 1983; *Sociologia Ruralis* 1988; Tulla 1989; Whatmore 1991), with very few exceptions this research has dealt with the family farm and farmwives while the study of female salaried agricultural workers in developed countries has been ignored (Henshall-Momsen 1989; Redclift and Whatmore 1990; García-Ramon and Cruz 1992). However, *latifundia* (large estates with hired workers), and not family farms, are the predominant type of production unit in several regions of southern Europe, such as the Italian Mezzogiorno, or Andalusia in Spain. Although outmigration and mechanization in recent decades have diminished their numbers, landless salaried workers (including female wage labourers) remain a significant feature of the social landscape of those regions. It is our hypothesis that the key issues that have proved relevant in research on women on the family farm (sexual division of labour, the invisibility of domestic work, patriarchal relations within the family) are also crucial for understanding the specific role that salaried women play within the family in the productive sphere as well as in the reproductive one.

In this chapter we also want to highlight some specific effects of welfare state policies on women's everyday lives (Gordon 1990; Lewis 1992), in particular in southern Europe where the welfare state has been relatively undeveloped in comparison with northern Europe. As Paola Vinay indicates with respect to Italy in Chapter 10, the welfare state in southern European countries very often chooses to provide people with financial assistance rather than social services, and the family has developed as an alternative protective net. This holds true for Spain where not only was the welfare state undeveloped but also it came late. It is clear that a number of indirect subsidies for low-income people that are quite common in northern countries (like public housing, public nurseries, facilities for elderly people, etc.) are lacking in Spain, particularly in the rural areas of the poorer regions of the country (Chiarello 1989). This is true even in Andalusia where the rural population lives in rather large agro-towns and not in scattered settlements. By contrast, since 1982 when the Socialist Party came into power, the welfare state is heavily present in the rural economy of these areas through policies of unemployment subsidies for casual daily labourers. These policies have unexpected and complex effects upon the relations between women's productive and reproductive work. In this chapter, therefore, to examine the ways in which women manipulate state policies, we will present a case study of day-labourers in the municipality of Osuna (close to Seville) which is fairly representative of agricultural communities.[1]

THE STUDY AREA AND THE FIELDWORK

Salaried work in Andalusian agriculture

Agricultural employment in Andalusia consists mainly of salaried work and, particularly, of casual work by day-labourers (Cruz 1987). The land ownership pattern has always been defined by the absence of medium-sized family farms and the overwhelming presence of large farms that are located on the more fertile soils. At the same time, there are numerous very small farms unable to sustain the owners' families. The large farms are mainly devoted to extensive dry farming, dominated by the cultivation of cereals and olive trees, except in some areas where highly specialized vineyards are important. In recent decades, the massive introduction of chemical fertilizers has brought about the suppression of fallow land, and new crops alternate with cereals – mainly sugar beet, cotton and sunflowers. The small number of crops concentrates agricultural employment in short periods and gives rise to widespread seasonal unemployment. Whereas day-labourers used to work on piece-rates (*a destajo*), they now work for a day wage (*a jornal*) negotiated by the unions.

These rural workers may complement their income as migrant workers by taking advantage of temporal variations in the harvest of one crop in different areas or by moving into areas specializing in different crops. This seasonal migration used to be done on a family basis as long as piece-rate payment allowed the workers to maximize income by employing all members of the family (including women and children) who could contribute in any way and thereby increase the final pay. It has often been asserted that female work was unimportant in the Andalusian family farm. Though this is true of family members in the few areas where there are family farms, day-labourers' wives and daughters have always been engaged in work in the fields. The practice has created a negative social image of women's involvement in agriculture in contrast with other regions of Spain where family farms are the dominant production unit and women are mostly farmwives.

Agricultural day-labourers and the welfare state

The agrarian unemployment subsidy policies of the socialist government in Andalusia have been very important for understanding the economic basis of Andalusian rural families (Mansvelt Beck 1988). Such policies – Subsidio Agrario and Plan de Empleo Rural (PER) – aim to complement the income of casual or seasonal agrarian day-labourers in both regions where they are quite prominent in the structure of agrarian employment. The Subsidio Agrario specifies that agricultural day-labourers who work 60 days per annum on a farm not belonging to their family are entitled to unemployment subsidy for 180 days at a rate equivalent to 75 per cent of the national minimum guaranteed wage (*salario mínimo interprofesional*), that is, 1,300 pesetas per day in 1990.[2] Complementary to this, the PER allocates large amounts of money to the local administration (mainly town councils) to fund public works programmes that are required to hire unemployed rural labour. On the one hand, PER makes it much easier for day-labourers to attain the 60 days' work target but, on the other, the policy serves as a powerful mechanism of 'political clientelism' for the local administration that has the responsibility for hiring the labour. A consequence of these policies is clearly reflected in the agrarian statistics of the 1980s which show an increasing registration of both the agrarian active population and agrarian unemployed population (Table 13.1), mostly accounted for by groups like women and young people (above 18 years old) who formerly did not register (for example, women amounted to 32.4 per cent of the total subsidized agricultural population in 1988 and by 1991 this share had risen to 51.4 per cent).[3]

The municipality of Osuna and the fieldwork

The area chosen for this study is the municipality of Osuna, fairly representative of Andalusian agro-towns with a well-developed urban structure. It is located 88 km from Seville on the main transport axis between this city and Malaga and Granada. Today Osuna has 16,728

Table 13.1 Agrarian active population and unemployment

	Agrarian active population	Percentage of unemployed
1981	431,000	18.1
1983	400,400	17.2
1985	474,400	32.5
1988	488,100	34.3

Source Boletin de Informacion Agraria y Pesquera, No. 27, 1990

inhabitants compared with over 24,000 in 1940 before the beginning of the intense migration process that affected the whole of rural Andalusia and lasted until 1980. According to the 1991 census, almost 50 per cent of the economically active population in Osuna is engaged in agriculture, 35 per cent in the services sector, 9.3 per cent in construction and only 4.1 per cent in industrial activities. Most (95 per cent) of Osuna's large municipal area (59,100 hectares) is devoted to agrarian production and especially to cereals (39,158 hectares) and olive trees (10,348 hectares). The pattern of land ownership is dominated by large farms above 100 hectares, which account for about 45 per cent of the total cultivated area.

Our research draws primarily on field data because from our previous studies (García-Ramon *et al.* 1990) we know that Spanish agrarian statistics are very poor for studying women's work. We completed fifty survey questionnaires and ten in-depth interviews. The female day-labourers interviewed were chosen from a random sample of a complete list of all the female labourers provided by the municipality of Osuna. The questionnaire consisted of four sections: the first referred to the composition and economic activity of the household; the second dealt with woman's work as a day-labourer in agriculture; the third referred to housework and the care of the children; the fourth included questions on management of the household budget, decision-making and leisure time.

To obtain more qualitative information we supplemented the survey with ten in-depth interviews of women who earlier had completed the questionnaire. At the present stage of knowledge of the subject it is hard to grasp some aspects of farm women's work and life, especially their attitudes and opinions, so that such interviews are important not only intrinsically but also to help us to widen our background and knowledge of the topic and to interpret the questionnaires. In the following sections of the chapter we will combine the results of the survey and the in-depth interviews. The quantitative comments come from the survey and are summarized in the tables; the qualitative analysis draws on the interviews.

CHARACTERISTICS OF THE FEMALE DAY-LABOURER HOUSEHOLDS

The households of the fifty women interviewed have a total membership of 258 persons, which means an average of 5.16 persons per household. Nuclear families are clearly dominant since female day-labourers with their husbands and children make up 85 per cent of the total population of the sample. Nevertheless, some households include two generations – usually when the second generation is a young couple – because housing is becoming very expensive. There are 122 children; that is, an average of 2.71 children per married woman worker.

The mean age of the sample population is 27.9 years (Table 13.2),

Table 13.2 Age structure of the survey population

	No.	Percentage
15 years	80	31.0
15–64 years	169	65.5
over 64 years	9	3.5
Total	258	100.0

Source Authors' fieldwork

reflecting the incidence of young children, and the mean age of the women interviewed is 40 years. This fact, and the dominance of the nuclear family, explains the low proportion of persons above 65 years (3.5 per cent) in this population. The level of education of the population is very low (Table 13.3) and it is even lower in the case of women workers: 48 per cent of them never went to school and of the remaining 52 per cent none went beyond the level of general basic education. Given the observed reluctance to acknowledge one's own illiteracy or low level of education, it may be surmised that such figures underestimate the real situation.

Within the sample a total of 149 persons are registered as economically active, indicating a high rate of activity (83.7 per cent) of the population above 15 years of age. Families report an average of 2.56 day-labourers each, reflecting their dependency on seasonal agrarian activity but also on a regular subsidized income. Overwhelmingly, families consist of day-labourers with 86 per cent (128 individuals) within the sample working in agriculture (Table 13.4). The figures demonstrate the lack of diversification of the economic basis of the families in the sample.

THE FEMALE DAY-LABOURER

Agricultural work and the sexual division of labour

As shown in Table 13.5, the age distribution of the women in the sample is relatively homogeneous, with a strong presence of women between the

Table 13.3 Level of education

	Total Sample %	Female day-labourers %
Without schooling	23.1	40.0
General basic education, 1st level	55.1	40.0
General basic education, 2nd level	15.8	12.0
Other	6.1	—

Source Authors' fieldwork

Table 13.4 Economic activity of household members

	Female day-labourers No.	Husband No.	Other members No.	Total No.
Day-labourers	50	35	43	128
Building industry	–	4	1	5
Industrial activities	–	1	2	3
Service sector	–	1	12	13
Total	50	41	58	149

Source Authors' fieldwork

age of 30 and 39 years (more than one third of the sample). The youngest woman is 22 years old and the oldest is 61. Most women are married and only five – the youngest ones – are single; there is only one separated woman in the sample.

According to the survey, all these women seek employment as day-labourers because of insufficient family income. They all assert that without their economic contribution the family could not survive. The need to work as day-labourers in order to ensure the survival of the family is also reflected in the answers concerning the age at which they started to work: the mean is the age of 13 years (Table 13.6). One woman reported that she had to start working at the age of 8, but most of them (two thirds) began working as day-labourers between 10 and 13 years of age. These women were able to work while they still were children because day-labourers often went to work seasonally in the fields with all the family regardless of sex or age. This practice was favoured by the type of contract (*a destajo*); it had obvious consequences in the low level of education already mentioned. The picture is quite different today; when the women are asked about it, they think that the situation of their children is better than was their own because the children can attend school much more often than they did and thus get some kind of instruction.

Today, except in a few cases, these women do not take their children with them to work in the fields. Most work is now contracted on a daily

Table 13.5 Age structure of female day-labourers

	No.	Percentage
21–9 years	11	22.0
30–9 years	18	36.0
40–54 years	15	30.0
55–64	6	12.0
Total	50	100.0

Source Authors' fieldwork

Table 13.6 Age at first employment of day-labourers

Age	No. of women	Percentage	Cumulative %
8	1	2.0	2.0
9	4	8.0	10.0
10	4	8.0	18.0
11	2	4.0	22.0
12	12	24.0	46.0
13	5	10.0	56.0
14	10	20.0	76.0
15	3	6.0	82.0
16	4	8.0	90.0
17	1	2.0	92.0
18	2	4.0	96.0
19	2	4.0	100.0

Source Authors' fieldwork

basis and not at a piece-rate; farmers are thus not interested in hiring children, and additionally, regulations concerning children's work are more strictly controlled. Moreover, the agrarian unemployment subsidies apply only to people over 18 years of age and children below this age cannot take advantage of them. As a consequence, children's work has decreased markedly.

An important feature of the work of these female day-labourers is its strong seasonality. Osuna's day-labourers work in only four crops: the *almazara* (or black) olive (to make olive oil), picked up in January and February; the *verdeo* (or green) olive (to be eaten as such), harvested in September and October; the strawberry, harvested from March to May; and the cotton, harvested in September. There are two major peaks of activity, one in January and February (37.8 per cent of the total work days registered) and a second in September and October (38.4 per cent). There is a less marked peak in March, April and May (15.9 per cent) and a period of virtual inactivity during the summer.

Picking olives still remains the major source of female employment in Osuna, accounting for almost three quarters of all work days. It was the traditional activity in this area and the difficulties of mechanization explain its contemporary relevance. Women say that they prefer picking the green olives to picking up the black ones because in early fall the temperature is good and the work can be done standing up. By contrast, picking up the black olives in late fall means that the weather is much worse and women have to bend down all day. Women complain that it is not easy for them to get work with the green olives since employers prefer men, both because they can hang the trees harder and also because they use the *banco* (a kind of ladder that is quite heavy) more easily. Employers prefer to hire women for picking up the black olives because this requires being bent all the time and they think that women do it much better; additionally, this task has undergone a certain process of feminization in the areas of olive groves as women, nowadays, do not want to move out of town (to get a better job) because of having the children in school, and therefore more women are available for picking up the black olives.

Picking strawberries, a minor source of employment for day-labourers (12.5 per cent in our sample), is an activity that makes the female workers of Osuna move 200–30 kilometres from their homes because it is located on the coast of Huelva province. The crop is not traditional, having been introduced only in the last ten years. But women do not like this work because they have to leave the town and somebody has to take care of their children (usually the mother or the mother-in-law). They seem to do it only if they have not been able to complete the 60 days required to apply for unemployment benefits. Again, this work is thought to be a woman's job but they complain that it is very tiring because they have to bend all the time and suffer considerable back-aches. Strawberry-picking is one of the few expanding activities with an increasing demand for seasonal labour and it is somewhat better paid. However, women in Osuna experience strong spatial restrictions in participating in it due to their traditional family responsibilities and the distance to the Huelva coast.

The cotton harvest now accounts for only 6.5 per cent of days worked by the landless people of Osuna. This represents a significant decline in importance over the last twenty years, as a result of increased mechanization of harvesting. Women do not like the task much as it also requires being bent all the time. It is also thought to be a 'woman's job' because you need nimble fingers to do it well, as employers say.

In only three cases of our sample have female day-labourers worked seasonally in the sunflower harvest, a very important crop in Andalusian agriculture but one where production is almost completely mechanized.

On average, the female day-labourers report 56 days of work per year and, in general, they have an optimum target of 60 days of work per year, which closely relates to the minimum of days required in the current legislation on subsidies for agricultural unemployment.

Although at the time of the survey women said they worked only as agricultural day-labourers, many of them (40 per cent) have reported formerly working in activities other than agriculture in the same town. Fifteen have worked as domestic servants and four in an olive oil processing plant. From the in-depth interviews however, we learned that quite a few women in fact supplement agricultural work with employment as domestic assistants in middle-class households in Osuna. Sometimes they work a few hours a week during the period that they do not work in the fields. Of course, this work is not declared as they are officially unemployed.

From the interviews we also discovered that undeclared textile work is becoming widespread in Osuna, mainly in co-operatives, that is, in small workshops. This activity started with women who are not agricultural day-labourers and some of the latter seem to envy such work. Yet they also claim that it requires too many hours regularly per day (about ten hours) and they cannot afford this time. We might, however, make the interpretation that they do not value this alternative because they can get the unemployment subsidy by working only 60 days and they are used to agricultural work since childhood.

With regard to salary, 51 per cent of our sample earned 2,600 pesetas a day and 70 per cent between 2,500 and 2,700 pesetas. Such a degree of uniformity is to be explained by the general incidence of collective wage bargaining in the countryside. On average, women day-labourers earn 145,600 pesetas a year to which has to be added another 234,000 pesetas of unemployment subsidy for those who have succeeded in working 60 days a year (if they have worked fewer days there is a lower subsidy).

THE WORKING DAY AND THE ROLE OF THE FEMALE DAY-LABOURER WITHIN THE FAMILY

Only nine women (out of forty-one valid answers) in our survey reported not receiving any kind of help from members of their family in tasks related to the maintenance and repair of the house, tasks that are clearly seen as within men's domain rather than women's. In other types of housework, the proportion of women saying that they never get any help from other members of the family is greater, though, on the whole, a majority of women still answer that they receive some kind of co-operation, either on a daily basis or more sporadically (at weekends, for instance). The extent of daily co-operation differs by task: 14 per cent of women are helped daily with childcare, 30 per cent with shopping, 41 per cent with cooking and 48 per cent with house maintenance.

We get a more detailed picture of the female day-labourer's working day from the in-depth interviews. The women really bear a double workload – in the productive sphere as well as in the reproductive one. When they work in the fields, the schedule is very tight. They usually get up at 6.30 or 7.00 a.m. to prepare breakfasts and lunches to take out. Then they go to the field and work from 9.00 a.m. to about 4.00 or 5.00 p.m. with a pause of an hour for lunch in the field. Afterwards they return home and do the housework – cleaning, making beds, preparing the *merienda* (some snacks for the children) and so on. Later, they prepare supper in order to have it about 9.00 p.m. On Sunday they do not work in the fields but spend more time on housework.

Despite this heavy time investment at home, it is clear that female day-labourers get more family assistance in reproductive tasks – even from their husband – than farmwives; previous research indicates that this holds true not only for Andalusian farmwives but also in other areas (García-Ramon and Cruz 1992). The attitude of the day-labourer's husband is very '*machista*' but there is no real alternative if the woman has to contribute to the family income (and we have to take into account that men's unemployment is quite widespread). For example, the husband does a few errands, especially shopping, when he comes home. But if possible, he prefers to go straight to the neighbourhood bar and stay there until supper time. It is interesting to point out that when women are asked about their husband's help with the housework they often state that in rural areas attitudes are quite backward and men do not contribute to housework tasks as they do in the cities.

Nevertheless, the real contribution to reproductive work comes from the mother (or the mother-in-law) and from the eldest daughter. When the children are very small, the woman's mother or mother-in-law usually takes care of the children during the periods when she is working in the fields. Depending on the circumstances (usually when the children are a little older) the eldest daughter (maybe 10 to 13 years old) takes care of them, and thus cannot attend school and has to drop out as her mother did. Even when they do not quit school, the daughters – and not the sons – help their mother a great deal with housework from a very early age. In the very few cases in which we find married women with children moving to work to the coast of Huelva, it is the mother and the mother-in-law who take care of the children when the female day-labourer is away.

With respect to leisure time, married women report that they never go out without their husbands, either because they are not used to going alone or because this is not well accepted in Osuna where it is quite traditional for men to meet and stay together in the bars but never for women. For women to go out by themselves is thought to be a modern habit, very typical of a large city but not at all of a town like Osuna. Actually they go out very seldom, even with their husbands (usually for the traditional festivals of the town), and none of them takes vacations. Nevertheless, it should not be thought that they do no social networking. They do it in a way that probably cannot be classified as leisure; for example, queuing up in the shops, waiting in the doctor's office, or going to church.

When asked about decision-making, the women in our survey emphasize that decisions are made collectively, especially answering that couples make decisions jointly (Table 13.7). But from a more detailed analysis of the in-depth interviews we observed a difference between communicating a decision and taking a collective position; many times the answer 'making a joint decision' really meant that the husband told them he was going to take such a decision. However, women are very active in making decisions about children and especially about their education.

According to the survey, in the vast majority of cases the women report that they spend their salary on basic maintenance of the house and on family needs. Only one woman reported that she spends the salary on

Table 13.7 Decision-making within the household*

	Husband No.	Wife No.	Couple No.	Family No.	Total No.
Children's education	–	7	37	3	47
Home improvements	2	3	38	6	49
Buying a car	–	3	42	5	50
Buying furniture	1	3	42	4	50
Buying electrical appliances	1	3	42	4	50
Total	4	19	201	22	246

Source Authors' fieldwork
Key * The answers were not exclusive

personal needs, another to pay back credits; in two cases the salary is put aside for savings. From the in-depth interviews, it is quite clear that their salary is basically spent on family needs but often on items related to the children, especially children's clothes, and the husband's salary is allocated to most basic things such as food. Single female day-labourers say that they keep the salary of the 60 days' work for themselves and give the unemployment subsidy (which amounts to almost double) to their family.

Women's perception of their own situation and perspectives for the future

When women were asked if they would like to abandon agriculture and engage in another job, forty-five out of fifty answered that they definitely would do so because agricultural work is too hard and unreliable. When asked about the type of work they would like, their answers showed an awareness of the limits imposed by their extremely low level of education and training – 39 per cent answered that they would work as house servants, 26 per cent as factory workers and 10 per cent as shop attendants. Four women declared that they would do any kind of work rather than agriculture. The five single women within our sample said that they would like to quit work if they got married, a desire that remains a wishful thought, as shown by the results of the survey.

Thus, we can see that most of these women would like to abandon their work as agricultural day-labourers if they had the opportunity to do so. This conclusion emerges even more strongly in the in-depth interviews. The reasons given are always the harshness and precariousness of the work. These families are in the lowest-income levels of Andalusian society and it is natural that the women wish to change their status. However, they are at the same time very aware of the tremendous difficulties of accomplishing that wish, due both to the lack of alternative employment locally or outside, since migration is no longer an outlet, and to their lack of vocational training and extremely low level of education.

What these women see very clearly is that they do not want this kind of life for their children and they express themselves that way in the interviews. One of the reasons they do not want to work out of town is because it would probably mean that the children would have to leave school. But reality seems to be somewhat harder. Regular schooling is quite complicated (especially for an eldest daughter) when the mother works in the fields. Further, in the few cases we have found of children studying away from home with a scholarship, they face quite significant difficulties. One daughter, for example, who was studying law in Seville had to work the first year in order to help her family with some money; the following year she had to go home for several months while her mother was sick.

CONCLUSIONS

It is clear that female day-labourers in Osuna work out of necessity in order to complement the meagre family income. These women are burdened with the problems of a double work day, the paid work and the domestic one. For a few months they are engaged in the fields as much as six to eight hours a day in addition to their normal household chores. During the rest of the year, quite a few women are employed for a few hours daily in paid domestic work for middle-class households and thus they also experience the double work day to some extent. Female day-labourers' husbands share household tasks a little more than farm-wives' husbands but they do it out of necessity because women's income is vital for the survival of the household. As soon as they can, they go to the bar with their male friends. Women complain about this situation but also say that they cannot train their sons in household chores in the same way as their daughters because of social pressures which they believe are stronger in the rural areas than in the cities. The daughters – and not the sons – of the female day-labourers help a great deal with housework; they have to take care of the children and quit school for a while if their mothers have to work in the fields and other female members of the family cannot take care of them. It is quite plausible that many will become school drop-outs. This lack of education also seems to be one of the reasons that they start working in the fields although they have to wait until they are 18 years old under present legal regulations.

The destination of women's salary and unemployment benefit is a clear indication of the subsidiary role it plays: the husband's income is spent on very basic needs (as women see it) like food and housing. By contrast, women's income is allocated to less important items, like children's needs or clothes, perhaps more directly related with the reproductive sphere. It is clear that in a patriarchal society – like the

Andalusian one – non-domestic production is the men's primary concern and women's involvement in it is viewed as secondary to their reproductive activities. Therefore, a woman's salary is regarded as complementary and not as the primary source of the family income.

One of the basic consequences of women's reproductive activities is restriction of their mobility, a feature prevalent in most societies though in different degrees of intensity. In Osuna the well-accepted social rule that women do not go out by themselves to public places like the bar or local festivals, and go only to places closely related to the reproductive sphere, such as shops and doctors' offices, might be interpreted from that perspective. But, above all, activities compatible with reproduction imply a low degree of physical mobility; thus, women's strongly felt restriction on taking a job out of town, like picking strawberries on the coast, is clearly a spatial limitation derived from their reproductive role because they do not want to have to take their children out of school. Some years ago this was not seen as a problem because education was not valued at all, but nowadays, at least in theory, education means the possibility of some upward social mobility. In Spain it has often been asserted, especially by the liberal wing of a wide array of parties, that given the difficulties of changing the economic structure of Andalusia and enlarging its employment possibilities, a more feasible target would be upgrading the level of instruction and vocational training (especially for women) to enhance their chances of social and economic promotion. But from a feminist perspective this would not be a sufficient solution as education, by itself, would not alter women's position, in that education cannot address issues of childcare and domestic work which clearly limit the full participation of women in non-home production and in the full appropriation of society's output.

In contrast with the assumption that the division of labour by sex is natural or given due to women's physiology and role in reproduction, feminist literature (for example, Benería 1979; Benería 1993) stresses the fact that it is a result of the subordination of women in our society – as in many others – and that it is subject to change. The case of female day-labourers in Osuna clearly exemplifies this situation. Fifteen years ago the salary of a female wage labourer and a male one for the same type of work was not the same. Nowadays, with the general application of collective wage bargaining in the countryside, women's salary and men's is the same for a given type of work. Nevertheless, we observe that women tend to concentrate in some jobs and that certain types of work have suffered a process of feminization (like picking black olives, strawberries and cotton but not the green olive). Through this mechanism they get paid less. The reasons for this are based on the different stereotypes of women's attributes, like nimble fingers for the cotton or resistance to staying on the ground and being bent for many hours for the black olives. But it is argued that these skills are natural and not acquired and thus are not paid extra (as if domestic activities were not an important training for nimble fingers and improving physical resistance).

It is clear that one important consequence of welfare state policies in Andalusia is that they have made women's work in agriculture more visible, especially the work of casual day-labourers. But it is also true that this has often made women's paid work socially devalued because the only important and necessary effort is to be able to gain credit for 60 days' work, many of them in often trivial subsidized public works. Surely, this is not the way towards a fully egalitarian participation of women in the labour market – even less towards the acquisition of higher qualifications and skills. In a sense, women more often than men tend to become somewhat the specialists in working for the Plan de Empleo Rural in order to get the subsidy to complement the meagre family income. This is quite clear when there is a choice to be made within the family unit.

Moreover, it should be pointed out that the welfare state has come later to Spain – as to other southern European countries – than to northern countries. It has developed mainly through unemployment subsidies rather than through indirect subsidies for low-income people (with the exception of basic medical care) such as the provision of public housing, public nurseries and care for elderly people. It is true that in many instances this deficit has been partially compensated by the family, which serves as a protective net. Thus support by the family is essential for understanding women's integration in the labour market, as is clear in the case of Osuna. For it is not families but specific members within them (mothers, mothers-in-law and eldest daughters) who often make it possible for the female day-labourer to work outside the home. This alternative protective net implies the existence and persistence of prescriptive behaviours; that is, certain rights as well as duties – as Dina Vaiou points out in Chapter 3. Therefore this situation might reinforce, in turn, the existing sexual division of labour and the patriarchal relations within the family. Thus if we want to change the current pattern of social reproduction, welfare state systems should be gender-aware, and planning policies should give particular attention to the gender system of rural societies.

Finally, our findings imply that the existing gender division of labour cannot be taken as given – even in agriculture – and this issue has to become a focal point around which far-reaching changes need to take place. How we must proceed in order to accomplish change is much more difficult than just stating the problem. As we have shown, even commonly proposed solutions such as improving the provision of education and fostering changes in educational aspirations will be of limited utility while the present pattern of social reproduction prevails. It has been said (Benería 1979) that an initial step in this direction is to analyse the situation in very different regional contexts in order to set up a general conceptual framework from which to proceed. It is to this effort that our chapter has been addressed.

NOTES

1 This paper is part of a wider research project funded by DGICYT No. PB 87-0769, DGICYT No. PB 90-0710 and DGICYT No. PB 93/0846. Besides the authors of this chapter, M. Baylina, A. Caballé, G. Cànoves, C. Domingo, I. Salamaña, M. Solsona, A. F. Tulla, N. Valdovinos, M. Vilarino and R. Viruela are participating in the wider research project.
2 In 1990, one US dollar was worth about one hundred pesetas.
3 This information comes from the *Boletin de Informacion Agraria y Pesquera*, December 1988 and April 1991. Unfortunately, the information given by sex is very limited and does not allow much further elaboration.

REFERENCES

Barthez, A.(1982) *Famille, travail et agriculture*, Paris: Economica.
Benería, L. (1979) 'Reproduction, production and the sexual division of labour', *Cambridge Journal of Economics* 3: 203–25.
—— (1993) 'Comptabilitzant el treball de les dones: una avaluació del progrés de dues dècades', *Documents d'Anàlisi Geogràfica* 22: 91–113.
Canoves, G., García-Ramón, M. D. and Solsona, M. (1989) 'Mujeres agricultoras: un trabajo invisible en las explotaciones familiares', *Revista de Estudios Agrosociales* 147: 45–70.
Chiarello, F. (1989) 'Sussidi, dualismi e discontinuità, in [assorted authors] *Storia d'Italia. e regioni dall'Unità a oggi*, Turin: Einaudi.
Cruz, J. (1987) 'Political and economic change in Spanish agriculture', *Antipode* 19, 2: 119–33.
García-Ramon, M. D. (1990) 'La divisón sexual del trabajo y el enfoque de género en el estudio de la agricultura de los países desarrollados', *Agricultura y Sociedad* 55: 251–77.
García-Ramon, M. D. and Cruz, J. (1992) 'Vita e lavoro delle donne nella Spagna rurale: lavoratrici giornaliere e la divisione sessuale del lavoro in Andalusia', *Inchiesta* 22, 96: 51–8.
García-Ramon, M. D., Solsona, M. and Valdovinos, N. (1990) 'The changing role of women in Spanish agriculture: analysis from the agriculture censuses, 1962–82'. *Journal of Women and Gender Studies* 1: 135–63.
Gordon, L. (ed.) (1990) *Women, the State and Welfare*, Madison: University of Wisconsin Press.
Haugen, M. S. (1990) 'Female farmers in Norwegian agriculture from traditional farm women to professional farmers', *Sociologia Ruralis* 30, 2: 197–209.
Henshall Momsen, J. (1909) 'Género y agricultura en Inglaterra', *Documents d'Anàlisi Geogràfica* 14: 115–30.
Lewis, J. (1992) 'Gender and the development of welfare regimes', *Journal of European Social Policy* 2, 3: 159–73.
Mansvelt Beck, J. (1988) *The Rise of a Subsidized Periphery in Spain: A Geographical Study of State and Market Relations in the Eastern Montes Orientales of Granada, 1930–1982*, Amsterdam: Selecta.
Redclift, N. and Whatmore, S. (1990) 'Household, consumption and livelihood: ideologies and issues in rural research', in T. Marsden *et al.* (eds) *Rural Restructuring: Global Processes and their Responses*, London: David Fulton Publishers.
Sachs, C. E. (1983) *The Invisible Farmers*, Totowa, NJ: Rowman & Allenheld.

Sociologia Ruralis (1988) monographic issue on 'Farm women in Europe', 28, 4.

Tulla, A. F. (1989) 'La mujer en las explotaciones agrarias del Pirineo catalán', *Documents d'Anàlisi Geogràfica* 14: 117–201.

Whatmore, S. (1991) *Farm Women: Gender, Work and Family Enterprise*, London: Macmillan.

14

WOMEN'S INTEGRATION INTO THE LABOUR MARKET AND RURAL INDUSTRIALIZATION IN SPAIN: GENDER RELATIONS AND THE GLOBAL ECONOMY

Ana Sabaté-Martinez

THE GLOBAL ECONOMY, RURAL INDUSTRY AND THE INCORPORATION OF WOMEN INTO THE LABOUR MARKET

Industrial decentralization in the 1980s: the global economy and the female workforce

The international division of labour in the global economy is giving rise to profound changes in the location of economic activities, as is well known. Industrial activity has been strongly influenced by these relocation processes especially in the labour-intensive industries which move in search of a cheap, abundant, flexible and non-unionized workforce (Dicken 1992).

In this economic situation, new regions (which were peripheral from an economic point of view) and new social groups (which were peripheral to the labour markets) have become very attractive to those light industries that use large numbers of unskilled workers. This is the case with clothing, toys, footwear and microelectronics manufacturing. The best-known cases of decentralization are certain peripheral cities and countries in the developing world, such as Singapore, Taiwan, Thailand, or Mexico. However, the same process has occurred in rural areas of southern Europe, although it is not so well known: this type of industrialization has been very important in northern Portugal (clothing manufacturing), in central Italy and in some Greek regions (clothing manufacturing spread in Anatoliki Macedonia in the 1970s, developed by firms from West Germany; Vaiou 1992). In Spain, the process was very important in the 1980s; we will analyse some aspects in central Spain, although this type of rural industrialization has spread to many other regions, especially near the Mediterranean (Houssel 1985). The whole process reflects the changing role of southern Europe in the new international division of labour (Hadjimichalis and Papamichos 1990).

Although the availability of a cheap and abundant workforce is the main reason for relocation, industrial geography has paid more attention to spatial and economic aspects than to labour markets. This scientific omission is the more surprising since the new labour markets have something in common: most of these industries overwhelmingly employ women as their workforce. As early as 1981 Helen Safa wrote:

> Few have given attention to the type of labor force recruited . . . Many studies do not even mention that the great majority of workers are women . . . The recruitment of women into these jobs is another stage in the search for cheap labor that characterizes industrial capitalism; these industries have a variable impact on the status of women employed by them in Third World countries.
>
> (Safa 1981: 419)

We can state that in Spain rural industrialization took place only in those areas in which there were big villages with a large population of young women without other employment alternatives (Sabaté 1989a). The incorporation of women into manufacturing in peripheral regions (like Taiwan, Thailand, Malaysia, or Mexico) is giving rise to completely new situations from the gender viewpoint, since these women have never had paid work before (Benería 1991); there are important similarities with the changes taking place in rural areas of southern Europe.

Rural industry and labour markets

The main industries to move to new sites in rural areas (southern Europe) have been clothing, textiles, footwear and leather, furniture and some electronics; all of them can fragment their productive process, all are labour-intensive and with the single exception of furniture manufacturing, all employ female workers.

Most of the decentralized industries are directly attracted by the advantages offered by rural labour markets:

- They can reduce salary costs by seeking low-paid workers: industrial salaries in rural areas have been estimated to range 15–20 per cent below those in urban areas.
- They do not need workers with any professional training, since only the most repetitive, boring, low-skilled and, therefore, low-paid tasks are decentralized.
- Many of these industries have a seasonal production rhythm, responding to market or seasonal demands (clothes, shoes, toys), which means that these industries use a temporary or seasonal workforce. Most of the food-processing industries also require seasonal workers and can be included here.

- The most effective way of reducing salary costs is to appeal to clandestine, temporary, home-based work, which always costs less than legal employment: it has been estimated that in some of these industries (footwear, leather, clothing) home-based work could range between 30 and 50 per cent of the total workforce (Sanchís 1987).
- The rural workforce is considered to be a 'docile' one, since these new workers do not have any experience with unions, and the unions themselves have no interest in small firms and dispersed workers. The personal relations between employer and employee are also determinant in avoiding any conflicts or strikes.

Female workers in rural areas offer all the advantages to decentralized businesses; they unite the condition of rurality to their being women – female 'docility' is very important in understanding the whole process. This means that female paid work is considered a 'simple' economic aid to the family, and, therefore, women's salaries are lower. Additionally, the social undervaluing of women's work means that women are willing to occupy temporary, unskilled, casual, home-based and low-paid jobs. Therefore, rural industry employs an overwhelmingly female workforce (like most decentralized industries) and the whole process of rural industrialization is deeply influenced by the gender divisions of labour and gender relations at the levels of family and society.

An emergent process: women's paid work in rural areas and changes in gender relations

As we have seen, rural industrialization has three main dimensions: regional, economic and gender. Our research has analysed the gender-related aspects at two different levels: 1) gender divisions of labour as the origin of rural industrialization; and 2) the changes that are taking place as a consequence of women's entry into the paid workforce.

The starting-point in rural areas of Spain in the 1970s could be considered 'traditional'. Gender roles were completely separated: the gender divisions of labour were very pronounced in rural areas; reproductive work and housework were the complete responsibility of women; and in rural areas women had always actively participated in productive work, although this fact has been socially and economically ignored.

This background deeply influences the conditions under which women enter the labour market in rural areas:

- Housework is still considered solely a women's chore. This implies that the productive work must be added to and combined with housework; rural women join the labour market at their own cost, taking on the double burden of domestic and paid work.
- Women in rural areas have not had any professional training and, therefore, can engage only in non-specialized jobs. Usually the only skills that women can sell on the labour market are those they have

learned at home, which are related to housework and caring (cooking, cleaning, sewing, taking care of children and the elderly).
- Both conditions make women's incorporation into the labour market very precarious: most women's paid jobs in rural areas can be classified as 'secondary', meaning that they are unstable, temporary, low-paid, unspecialized, clandestine and/or home-based.

Despite the precarious working conditions, this new role is introducing important changes in gender relations, changes which should be interpreted within the framework of the profound changes in gender relations that have taken place in Spain in the last twenty years. Among the changes that can be observed in the industrialized rural areas are new family strategies towards productive and reproductive work, a dramatic fall in fertility rates, changes in gender roles and the emerging of new gender relationships.

RURAL AREAS IN CENTRAL SPAIN: REGIONAL BACKGROUND OF THE RESEARCH

Rural change and family strategies

Most rural areas in Spain have suffered dramatic changes in the last decades: agricultural modernization has occurred abruptly through rural emigration, which was highest between 1950 and 1975. Before this process was completely accomplished, Spain's entry into the European Community and the world-wide organization of agricultural trade brought about a new restructuring of the sector with the main purposes of reducing the active population and the number of non-competitive farms. Since the 1980s, the actual income in most agricultural households has been decreasing and it will probably be even more reduced in the near future.

As in other European countries, economic diversification is being promoted to decrease the social effects of this reorganization (Arkleton Research 1990; Fuller 1990). The growing *pluriactivity* in rural areas usually implies paid work by one or more family members in a non-agricultural sector in order to supply a cash income that may permit the economic survival of the family farm. The process mostly affects young people and women (therefore, particularly, young women) who constitute an important new group of population that is looking for paid jobs. Female work is acquiring a new value in this context: women are going from unknown, unpaid and unrecognized activities to carrying out different salaried jobs, mainly in non-agricultural sectors.

Regional differences and opportunities of paid work for women

The differences in this general process vary markedly with the woman's age, the family's socioeconomic level and the geographical characteristics of the rural area. The individual situation within the family is very important, and especially age: at present, most young single women in rural areas attempt to get a paid job where they live, but not in agriculture, and they reject emigration to urban areas (which is completely different from what happened thirty years ago; Sabaté 1989b). Married women, with family responsibilities, are also taking on paid jobs to increase household income and mainly to 'complement' their husbands' wages. Socioeconomic and spatial variations introduce regional differences, and the size of the villages, land ownership, farm size, intensity of past emigration, proximity to cities, touristic resources or industry are all important.

Throughout Spain, local opportunities for paid work in rural areas are quite limited for women; the number and type of activities that they can perform are very scarce, and differ in different areas (Sabaté 1992); these are some of the main alternatives:

- There is temporary hired work in intensive and irrigated agriculture.
- There are jobs in the services (including work in hotels, as cleaning women and domestic helpers). Working in touristic areas usually implies temporary migration to the coast, as in Andalusia or the Balearic Islands; however, rural tourism is being politically encouraged in most of the rural areas in Spain and creates new opportunities for women in their own villages. Care of the elderly is also a new activity which is increasing in rural areas and which always employs women. It is important to stress that the jobs women take in services are jobs doing the same activities that they have always done at home, in an unpaid and unrecognized way.
- Rural industry can locally be the sector to create the most job opportunities for women. One of the most important aspects from the viewpoint of location is that industry exists where there are no other paid work alternatives for women (for instance, as we shall show, rural industry is never located near touristic or intensive agriculture areas).

Analysis of an investigation in central Spain

Within this general framework we have carried out a research project to analyse the relationships between the access of women to paid work and economic diversification in rural areas (via industrialization) (Sabaté 1990). We have selected a wide area in central western Spain, on both sides of the Sistema Central mountain range, including the provinces of

Salamanca, Avila, Segovia, Toledo and Cáceres (Figure 14.1); these provinces are part of three different autonomous communities (Castilla-León, Castilla-La Mancha and Extremadura), each of which has different regional economic policies. We have previously done a similar investigation in rural areas in the Autonomous Community of Madrid (Sabaté *et al.* 1991); its main conclusions are also incorporated into this chapter.

The area includes environments with different geographical characteristics: there are regions with small and large settlements, specialized and non-specialized agriculture, depopulated mountainous areas, etc. In general, these areas have suffered severe emigration, the agricultural returns are low and, specifically, *they have no prior manufacturing tradition.* The northern provinces (Salamanca, Avila and Segovia) have very small rural villages (150–200 inhabitants), they are mountainous, and their non-specialized agriculture produces very low revenues; emigration has been very severe in the last decades and there are very few young women in most of the small villages.

By contrast, the southern provinces (Toledo and Cáceres, like most of the La Mancha and Extremadura regions) have larger villages (1,000–4,000 inhabitants), these include areas of specialized agriculture and the population structure is more equilibrated. There are important irrigated areas in Extremadura, around the Tiétar, Alagón and Guadi-

AV=Avila (Castilla–León)
SA=Salamanca (Castilla–León)
SG=Segovia (Castilla–León)
CC=Cáceres (Extremadura)
TO=Toledo (Castilla–La Mancha)
M=Madrid (Community of Madrid)

Figure 14.1 Rural industry in central Spain: the research area

ana rivers, where crops like tomatoes, peppers, asparagus, tobacco and cherries locally create a large amount of seasonal hired work, for both men and women.

An area nearer to Madrid, about 100 km from the city, in the province of Toledo, has also been included; industrialization there was very important in the 1970s and 1980s. Examination of this area will allow us to compare the process of women taking on paid work in rural regions with different levels and rhythms of industrialization.

The investigation has two main levels of analysis, with a special emphasis on qualitative methods. First we identified the location of the industries in rural areas through conventional statistics; for this purpose we used the Spanish Directory of Industry (MINER 1989), which reports industry on the local level. It supplied information on the location, the main characteristics of the firms and the sectors of activity. Most of the research, however, is based on qualitative methods; using the statistical data we chose some villages and firms for the qualitative research. We performed 200 in-depth interviews, mainly with female workers, but also with some employers; the fieldwork took place in 1992 and 1993. The objective was to get information on the characteristics of the female workers, their working conditions, their family strategies to diversify incomes, the importance of casual work (temporary, home-based), the role that the gender division of labour played in the process, the double workload of women's productive and reproductive activity and the changes that may be taking place in gender relations. Employers were also interviewed in order to know the reasons for their locating in rural areas and their choice of women as a workforce.

ANALYSIS OF RESULTS

Location, characteristics of manufacturing and organization of production

In general, the industry was scanty and non-diversified, as is usual in deeply rural areas; however, the industrialization process has been very dynamic in the 1980s and has introduced marked regional differences (Figure 14.2). Although we have considered only firms with over five employees, it is clear that rural industry is quite important in the southern provinces (especially Toledo), and very scarce in the northern ones.

One of the clearest characteristics is the specialization in those sectors which are labour-intensive and require little capital and less technology; their cost reduction is made through the use of a cheap labour force. Table 14.1 summarizes the quantitative results for the whole area: 65.67 per cent of total employment (and 65.16 per cent of firms) correspond to only six sectors which are labour-intensive industries; moreover, five of these sectors are industries that overwhelmingly employ women as their

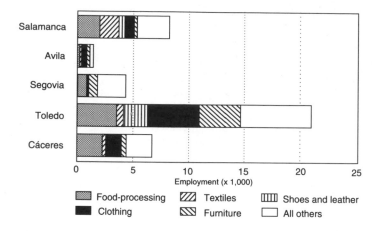

Figure 14.2 Rural industry in central Spain: total employment
Source MINER, 1989

labour force, and they represent some 52.05 per cent of the total employment in these regions. These five 'female' sectors are clothing (which includes 17.33 per cent of total employment), food-processing, footwear, leather and textiles. The only well-developed 'male' sector in rural areas is wood and furniture-manufacturing, which accounts for 13.62 per cent of total employment; if we compare these results from the five provinces with the structure of industry in Spain, the differences are very clear.

Table 14.1 Rural industrial structure in central Spain (five provinces)

Manufacturing sectors	Total units	Total employment	Total units %	Total employment %	Total employment Spain* %
Food-processing	463	8,751	20.76	21.21	14.78
Textiles	106	2,611	4.75	6.33	4.84
Leather	18	321	0.82	0.78	1.12
Footwear	133	2,638	5.96	6.40	–
Clothing	360	7,150	16.14	17.33	9.33
Furniture	373	5,618	16.73	13.62	8.63
Subtotal (for the above six sectors)	1,453	27,089	65.16	65.67	38.70
Other sectors	777	14,158	34.84	34.33	61.30
Total manufacturing	2,230	41,247	100.00	100.00	100.00

Key * Encuesta de Poblacion Activa 1993
Sources MINER 1989 and EPA 1993

There are regional differences, as is shown in Figure 14.2. In the three northern provinces the industry is diversified across a wider number of sectors and, in particular, the weight of the labour-intensive sectors is very low. The group for 'others' is very important and incorporates the traditional rural industries, like quarrying and mining (Salamanca, Segovia); the 'female' sectors are almost non-existent, especially clothing. By comparison, the southern provinces (Toledo and Cáceres) have a strong showing by 'female' manufacturing sectors, especially clothing, footwear (Toledo) and food-processing. The fieldwork (1992–3) has demonstrated that there were many more small firms than the ones reported in 1989 and, therefore, more employment.

The rural industries we analysed have a number of distinctive characteristics. First, with the sole exception of food-processing, the *size* of the firms is very small; the average clothing firm employs twenty workers. Second, the *organization* of these small firms involves *subcontracting* with larger enterprises that are located in the main cities of Spain (Madrid in this case): this is the usual way of organizing in a process of decentralization. As was said before, decentralization affects only the non-specialized, labour-intensive, low-paid tasks in industries with a limited need for capital. Third, there is a *high specialization and concentration* in specific areas: this means that most of the activities are not evenly spread over rural areas but concentrated in a few villages; for instance, textiles are located only in some small towns (Béjar, Sonseca, Plasencia); shoe-making is also very concentrated (Fuensalida, a village in the province of Toledo, offers 2,080 out of the total declared number of 2,638 jobs in shoe-making); furniture, which is the only 'male' sector, also has many workers in the province of Toledo, but they are concentrated in a small number of villages.

The *clothing industry* is probably the most interesting sector in rural areas: it is the one that has grown the most in the 1980s and it is also the one related directly to women's work: more than 90 per cent of the jobs are held by women. Its location differs from the other sectors in that it is more widespread; however, the small clothing workshops (or sweat shops) are always located in densely populated areas (where there is a good supply of labour force), in medium-to-large villages (concentrated settlement) and in places that lack other opportunities for women to work. These factors explain the difference between the northern and southern provinces: the clothing industry is not found in areas with small, depopulated villages (mountain areas of Castille), nor in places where there are other paid jobs for women (areas with irrigated agriculture or food-processing industries).

The organization of this sector is also typical of decentralized industries: the productive process is highly fragmented, so that only the cheapest and most labour-intensive tasks (sewing, ironing) are done in rural areas. One of the most interesting aspects is the important participation of *co-operatives* (Table 14.2), which is higher than in any other manufacturing sector

Table 14.2 Structure of the clothing industry

	Workshops	Employees	Average size
Co-operatives	145	2,922	20.1
Private firms	215	4,228	19.6
Total	360	7,150	39.7

(Figure 14.3). The size of co-operatives and of private firms is quite similar (twenty employees per workshop), and their organization is much the same: all work for big enterprises from urban areas (most are the big department stores in Madrid) through a complicated chain of subcontracting relations. The workers are overwhelmingly women (more than 90 per cent) and the salaries and other working conditions are quite similar in both co-operatives and private firms, although there are important differences between co-operatives. The location is slightly different: the highest concentration of co-operatives is in Toledo and Cáceres. The interviews show that the co-operatives have grown to give an opportunity for a paid job to women where there are no other alternatives. Most of them were established in the mid-1980s, when there were no working opportunities in the urban areas and when the restructuring of industry was most severe in Spain in response to the economic crisis in the 1970s and later in the global economy.

There are other regional differences which show that when workshops are located near Madrid (Toledo province) they can deal directly with

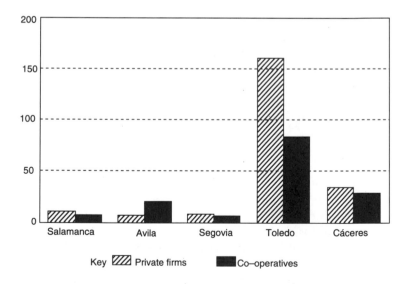

Figure 14.3 Structure of clothing industries in central Spain

the department stores, and therefore their salaries and working conditions are quite good. The further one goes from Madrid, the more complicated the subcontracting system becomes, involving intermediaries, and thus the worker's wages become lower, the workers are less specialized and, in general, the working conditions worsen.

A very important aspect of rural industrialization is the relationship between the clothing industry and *home-work*: this sector is notorious for its intensive use of clandestine workers, who are always women (Narotzky 1988; Sánchez-López *et al.* 1984; Sanchís 1987). However, in general, this type of working at home, or home-work, is not widespread in the research area and no clandestine work exists in the villages where co-operative organizations are dominant.

The interviews have demonstrated that there was a tradition of women sewing at home for the market in most of the villages in the southern provinces (this means informal and home-based work, paid on a piece-rate). This has been recorded especially in the La Mancha region, where the mechanization of cereal crops left women in villages without any job in agriculture. Additionally, the interviews showed that most of the co-operatives were established in the 1980s to improve the conditions of the women who worked at home in the submerged economy.

By contrast, the Spanish labour legislation in the mid-1980s gave fiscal advantages for setting up firms, especially small ones, thus allowing the growth of small private firms and fostering a decline in home-based work. Most of the decentralization process was based on the tradition and skill of women who had previously been sewing in isolation at their own homes.

The amount of home-based work today is quite unstable and its location reveals an irregular pattern; it is associated only with private firms. All the workers in this sector are women who are paid for piece-work and have an irregular and temporary work rhythm. Clothing manufacturers are thus able to maintain a seasonal production rhythm at the cheapest cost; we will see that there are important personal and economic differences between the women who work in the two sectors (formal and informal).

The location and characteristics of the food-processing industries are entirely different from the other sectors: the most important enterprises deal with canning (tomatoes, asparagus, fruits) and tobacco; they are therefore located near the areas of irrigated agriculture, in Extremadura, and nearly 90 per cent of their workforce is female. The work is only temporary, lasting from three to six months; the few permanent posts are usually occupied by men. The location of these activities is more related to the supply of products than to the labour market. There are other food-processing industries, however, in which location is independent of product supply; one peculiar specialization is the marzipan industry in Toledo, which is active three months per year, before Christmas time;

the large number of temporary jobs in this case are also entirely occupied by women.

Outside our research area are many other examples of food-processing industries located in rural areas: they may be independent of the product supply, but they are directly related to the availability of a large female labour force that is willing to work on a temporary basis. A good example is the production of most Christmas sweets: for instance, the village of Estepa (province of Cordoba, Andalusia) is well known as the centre of the processing of '*mantecados*': large and small enterprises employ 2,000 to 3,000 women for three months every year.

Most food-processing industries in rural areas are distinctive in comparison with other sectors: they are older (20–5 years old) and bigger; they usually are organized in a few large enterprises (many related to multinationals) and offer from 200 to 300 jobs per year in each site; most of the jobs are temporary ones held by women.

As we have seen, most rural industries take advantage of a cheap, flexible, 'concealed' and temporary workforce which allows them to reduce costs (through low salaries, flexibility and informality) and to adapt employment to seasonal production rhythms (through temporary jobs). It is evident that women so fulfil all the employment needs of this type of industry that they are one of the most important factors in choosing a location. Speaking about the characteristics of rural industrialization and of paid work for women in rural areas is speaking about quite the same thing.

Characteristics of female workers and their working conditions

The characteristics of all these rural women and their working conditions are very similar, although there are certain differences between the earliest and most industrialized areas (like Toledo province, near Madrid; Sabaté 1993) and the deep rural areas. In the *least and/or most recently industrialized areas* most of the women working in clothing workshops are very young (in some areas we found an average age of 20 years old; Sabaté *et al.* 1991), single, living with their parents, and without family responsibilities; they have received only a basic education up to 13–14 years old, and they start working without any professional training at the minimum age (16 years old).

The main reason these young women give for accepting this situation (despite the bad working conditions) is that 'there are no other alternatives to working' and they refuse to emigrate to urban areas. Usually they want to earn their own money and reject staying at home, doing housework, because they do not want to live 'as their mothers did'. By contrast, most of the home-work is performed by married middle-aged women, with children, who usually value this home-work positively, because 'it is easier for them to combine it with housework'.

The interviews held in the *most and/or earliest industrialized areas* (province of Toledo) revealed different patterns. The incorporation of women into paid work occurred quite early (it began approximately fifteen to twenty years ago) and there are more local opportunities for working in industry. The profile of the workers is therefore quite different: most of the women work in the formal sector, they are married, are middle-aged (30–40 years old) and have children. As we have already seen, home-work is quite rare in these areas, so that there is no segregation of women according to age between the formal/informal sectors. These conditions can also be seen in the food-processing industries where the women have been working for the last twenty years (in the irrigated areas in Cáceres, Extremadura).

All the responses we recorded in these second areas make it clear that a deep change is taking place, similar to the one that occurred with the incorporation of women into the labour market in urban areas: women work not only for economic reasons but because 'it makes them more independent', they have a clear idea that their contribution to the family income is very important (and not a 'simple aid' to their husband's income) and they prove that gender roles at the family level have begun to change.

The *working conditions* are very similar in all regions and manufacturing sectors: women are ignorant of their own rights as workers and even of their own situation and type of contract; there is a high frequency of irregular or alegal agreements (jobs may be temporary, discontinuous, or apprenticeships, though the actual working situation is different); the salaries are low but comparable with the official minimum salary at the lowest professional level. The low salaries are always justified on the basis of these women's lack of professional training. It is important to stress the very low activity of labour unions in these areas; this can be explained by the reduced size of most of the firms, and, in some cases, by the direct pressure of enterprises who threaten not to rehire any woman who has a link with a union.

Home-based and informal work mainly appears in the clothing industry and is concentrated in peripheral areas (like the border near Portugal, in Extremadura) and among 'peripheral' women (middle-aged, widows, or women with family problems, for instance); these women are given only the least specialized tasks in the clothing industry, which require no machinery.

Our conclusion is that most of the earliest clandestine work has changed into full formal work carried out in small firms and co-operatives; at least this was true in our research area at the beginning of the 1990s. In addition, we see that the *temporary regime* is another important characteristic of women's work, and it is the most common labour relation in the food-processing industries; to explain some women's preference for temporary conditions in paid work it is necessary to remember that women also do all the housework by themselves.

Housework, daily life and family

One of the most important aspects that we intend to clarify is the way in which women with paid work combine it with the housework and family responsibilities. The interviews have shown that, in general, housework is still only the women's responsibility. Against this background, temporary and home-based work is often considered by women to be the 'most appropriate', since it allows them to combine both productive and reproductive work; they consider that they can better do 'their' housework, either because they work at home (home-workers) or because they are out only an average of three months a year (seasonal workers). In the case of home-workers the situation is especially hard, because they must add housework to the seven–eight hours a day they already work; so, despite their answers, the only advantage that we can find is that they perform the two activities in the same place.

Nevertheless, it is important to stress some internal differences on the regional and the personal levels. Some subtle elements of change can be found only in the earliest and most industrialized areas, like Toledo province: in these places married women think that housework should be shared by both men and women, although in practice only a few husbands do some tasks like taking the children to and from school, reheating food, or shopping on the weekends.

Most of the young women in all the areas also have a clear idea about the need to share housework, since both men and women also bring money to the household; but, in practice, only some men are willing to 'give a hand to their wives'. In general, there is no doubt from the 200 interviews that women are taking on paid work at their own cost because they continue to support the whole weight of the housework.

The next question is: how do women manage to combine both types of work? Since we are dealing with rural areas, it must be remembered that public social services are generally scarce. The answer seems to be *the female family networks*: although families are nuclear in the research area, the proximity of the housing makes it easy to maintain close relations between family members. In the absence of most public social services, married women with children can take on paid jobs because of support by other women, especially their own mothers; mothers-in-law and sisters are also good helpers.

As Vaiou writes of southern Europe at the beginning of the 1990s:

> The traditional extended family has also been modified . . . however, it persists in new forms, where, for example, elderly parents and children's family may not share the same house but choose to live close to each other and share domestic and caring responsibilities: the elderly (more specifically grandmothers) look after young children and housekeeping, their daughters (or daughters-in-law) look after them when they need care.
>
> (Vaiou *et al.* 1991: 20)

This description (taken from examples in Greece, Italy, Portugal and Spain) fits the rural families that we have found in central Spain perfectly, and we consider it to be one of the most important features in Mediterranean countries that explains the incorporation of women into paid work.

The difficulties of holding a paid job, doing the housework and taking care of the children are producing a drastic fall in fertility rates among young female workers in rural areas, as has also happened in urban areas in Spain.

Finally, we should stress that there is a clear correlation between reproductive work, personal characteristics and working conditions: the more reproductive work a woman has, the smaller her opportunity to choose between different paid jobs: middle-aged women, with children, low incomes and living in peripheral regions, are left with the worst working conditions.

INTERPRETATION AND CONCLUSIONS

Feminist geography and industrial geography

The previous analysis is a good example of how feminist geography can improve the traditional interpretation of certain decentralization processes. It is necessary to introduce gender relations to explain the advantages that women offer decentralized firms. The undervaluation of women's work, their lack of professional training and, even more, their double burden of productive and reproductive work are some of the socially constructed gender differences that push women into temporary and less skilled jobs with low wages and, frequently, into the underground economy.

Thus the flexibility of women's work all over the world is a central feature that can help to explain the entire process of decentralization and flexible production, as we have seen in the rural areas of southern Europe. The economic crisis in the 1990s is revealing the fragility of rural industrialization: the search for an even cheaper labour force is moving many clothing workshops away to other countries like Morocco, at the same time as the market share of clothing made in India, Korea, Taiwan, or China is increasing.

Women as a 'captive' workforce: regional differences

The worsening economic evolution of rural areas in Spain compels women to search for paid work, because most households need more money. Most modern jobs in rural areas are 'not suitable' for women (agriculture, building, driving), so there are many less employment opportunities for women than for men (that is one of the reasons why women emigrated at a higher rate in the past decades from most of rural areas in Spain).

We have found that rural industries have developed only where women have no alternative employment: rural industry is thus non-existent in touristic regions, like the coastal Mediterranean, or the Balearic or Canary Islands. The rise in rural tourism has also generated more working opportunities for women and, therefore, it never coexists with industry in the same areas.

Women also have strong constraints on their paid jobs because of the burden of domestic work; even more, their accessibility to work is very limited (they drive less than men, who have the use of the family vehicle). Consequently, the lack of spare time and the low accessibility hamper women's access to distant jobs. This 'captivity' or immobility of the female workforce (Wekerle and Rutherford 1989) is crucial in explaining industrialization in rural areas: women offer good advantages to labour-intensive and temporary industries, since they need a wage but cannot negotiate their working conditions.

The location of rural industry is directly related to: 1) densely populated areas, where female emigration has not been very severe in recent years; and 2) the lack of other employment opportunities for women (like tourism and services in general). This is the situation that we found in the most industrialized areas (like Toledo province) and the marginal areas that also have large villages (most of Extremadura and La Mancha regions). By comparison, less populated regions have only a few industrial firms, reinforcing the lack of alternatives for women and leaving emigration as the only solution; this is the case of the northern provinces in our analysis (Segovia, Avila, Salamanca).

Family structures and gender relations in the new situation

Family structures are very important to explain the whole process; but they themselves are subject to changes derived from the new situation. First, the incorporation of women into salaried work is possible due to the persistence of *family networks* that are still strong and well structured in rural areas. These networks are very important because they fill in for the weakness or lack of state-provided social services, which should have facilitated the incorporation of women to paid work. The networks operate mainly among women belonging to the same family; the best example is the care of young children in a regional background (rural areas) where day-nurseries are very limited or non-existent: young women leave children with their own mothers, sisters, or mothers-in-law while they work at the workshops.

From the in-depth interviews another important distinction emerges among women: the differences between women derived from their age and *stage in the life-cycle*. The biggest differences in working conditions between men and women develop among the middle-aged, between 30 and 40 years old, when married women bear the heaviest weight of reproductive work; if they are in the labour force, most of them take on the worst positions, like temporary jobs (food-processing), home-based and informal work; the same can be said of women who are the heads of their households, like widows. So the process introduces new divisions between women in the labour market according to their age, children and life-course.

The incorporation of women into paid work is introducing certain changes in *gender relations*, as many in-depth interviews have made clear. Women value their work highly, not only because of economic advantages, but particularly because of its social and individual advantages (relations with other people, independence); this valuation is specially important in the youngest and unmarried women. As a consequence, gender relations have begun to change: no longer is the male partner the only breadwinner and, in turn, women are coming to hold the view that housework should be shared by men and women. This subtle change in gender relations is clear in the earliest and most industrialized rural areas near Madrid (Toledo province); however, we wish to stress that this change should be understood against the background of the deeper changes that have taken place in the entire Spanish society. It is very probable that these changes in gender relations are due as much to general factors as to local ones.

Reproduction costs, women's work and rural industrialization

The qualitative analysis of women's salaried work has revealed an aspect that is usually 'hidden' in studies about decentralization processes: reproduction costs are lower in rural areas, thus facilitating the lower wages (for both men and women): housing is cheaper, there are no expenses for transportation, all the meals are prepared at home, there are still basic products from the subsistence agriculture (like poultry, vegetables and olive oil) and many of the social services take place at home (care of children and elderly, meals preparation, etc.).

Feminist theory (McDowell 1991) and our own empirical analysis have made two aspects clear: 1) household networks are crucial to reducing the cost of social reproduction; and 2) the lower reproduction costs depend upon women's reproductive work. As we have seen, in rural areas, family networks among women work very well and women handle most of the reproductive work; thus, the lower reproduction costs found in rural areas basically depend on women. These advantages are being lost in southern Europe, since women are taking paid work outside their homes and the extended family is being replaced by the nuclear family (although in the conditions defined by Vaiou *et al.* 1991; see above).

In our opinion this is one of the main differences that can be observed in southern Europe (not only in rural areas, but also in small cities; see

Vinay 1985 for central Italy and Vaiou 1992 for Greece): the importance of family networks can explain the women's entering the labour market despite the weakness of public social services; these networks also explain the lower wages in rural areas (both male and female) and, in general, the shift of many firms from northern to southern Europe during the 1970s and 1980s. As the Mediterranean extended family tends to disintegrate, the advantages of these labour markets will move to the developing countries (as has already begun in the early 1990s).

REFERENCES

Arkleton Research (1990) *Cambio Rural en Europa,* Madrid: Ministerio de Agricultura.

Benería, L. (1991) 'La globalización de la economía y el trabajo de las mujeres', *Revista de Economía Sociología del Trabajo* 13/14: 23–35.

Dicken, P. (1992) *Global Shift: The Internationalization of Economic Activity,* 2nd edn, London: Paul Chapman.

Encuesta de Población Activa (EPA) (1993), Madrid: Instituto Nacional de Estadistica.

Fuller, A. M. (1990) 'From part-time farming to pluriactivity: a decade of change in rural Europe', *Journal of Rural Studies* 6, 4: 361–73.

Hadmichalis, C. and Papamichos, N. (1990) 'Local development in southern Europe: towards a new mythology', *Antipode* 22, 3: 181–210.

Houssel, J. P. (1985) *De la Industria Rural a la Economía Sumergida,* Valencia: Institución Alfonso El Magnánimo.

McDowell, L. (1991) 'Life without father and Ford: the new gender order of post-Fordism', *Transactions of the Institute of British Geographers* 16, 4: 400–19.

MINER (1989) *Registro Industrial de España,* Madrid: Ministerio de Industria y Energía.

Narotzky, S. (1988) *Trabajar en Familia: Mujeres, Hogares y Talleres,* Valencia: Institución Alfonso El Magnánimo.

Sabaté, A. (1989a) 'Geografia y género en el medio rural: algunas líneas de análisis', *Documents d' Anàlisi Geogràfica* 14: 131–47.

—— (1989b) *Las mujeres en el Medio Rural,* Madrid: Ministerio de Asuntos Sociales-Instituto de la Mujer.

—— (co-ord.) (1990) *Mercado de trabajo e industrialización rural: el recurso al trabajo de las mujeres,* Proyecto de Investigación subvencionado por la C.I.C.Y.T, no. PBS90–0563.

—— (1992) 'La participación de las mujeres en la dinámica social de zonas rurales desfavorecidas', in Ministerio de Obras Públicas y Transportes, *Desarrollo local y Medio Ambiente en Zonas Desfavorecidas,* Madrid: MOPT.

—— (1993) 'Industria rural en Toledo: la incorporación de las mujeres al mercado de trabajo', *Anales de Geografía de la Universidad Complutense* 12: 277–88.

Sabaté, A., Martín-Caro, J. L., Martín-Gil, F. and Rodríguez, J. (1991) 'Gender divisions of labour and economic change: the clothing industry in rural areas (Community of Madrid)', *Iberian Studies* 20, 1 and 2: 135–54.

Safa, H. I. (1981) 'Runaway shops and female employment: the search for cheap labor', *Signs* 7, 2: 418–33.

Sánchez-López, A., García, F., Ortiz, M. and Ruiz, M. C. (1984) 'La industria de la confección en zonas rurales del sur de Córdoba', *Estudios Territoriales* 13–14: 47–64.

Sanchís, E. (1987) *El Trabajo a Domicilio en el País Valenciano,* Madrid: Ministerio de Cultura, Instituto de la Mujer.

Vaiou, D. (1992) 'Il laboro femminile nel Mercato Unico europeo: un caso nel sud Europa', *Inchiesta* 96 (Aprile/Giugno): 44–50.

Vaiou, D., Georgiou, Z. and Stratigaki, M. (co-ords) (1991) *Women of the South in European Intergration: Problems and Prospects*, Brussels: Commission of the European Communities.

Vinay, P. (1985) 'Family life cycle and the informal economy in central Italy', *International Journal of Urban and Regional Research* 9: 82–97.

Wekerle, G. and Rutherford, B. (1989) 'The mobility of capital and the immobility of female labor: responses to economic restructuring', in J. Wolch and M. Dear (eds) *The Power of Geography*, Boston: Unwin Hyman.

INDEX